Designing the British Post-War Home

T0227467

In *Designing the British Post-War Home* Fiona Fisher explores the development of modern domestic architecture in Britain through a detailed study of the work of the successful Surrey-based architectural practice of Kenneth Wood. Wood's firm is representative of a geographically distinct category of post-war architectural and design practice – that of the small private practice that flourished in Britain's expanding suburbs after the removal of wartime building restrictions. Such firms, which played a significant role in the development of British domestic design, are currently under-represented within architectural histories of the period.

The private house represents an important site in which new spatial, material and aesthetic parameters for modern living were defined after the Second World War. Within a British context, the architect-designed private house remained a 'vehicle for the investigation of architectural ideas' by second generation modernist architects and designers.

Through a series of case study houses, designed by Wood's firm, the book reconsiders the progress of modern domestic architecture in Britain and demonstrates the ways in which architectural discourse and practice intersected with the experience, performance and representation of domestic modernity in post-war Britain.

Fiona Fisher, Faculty of Art, Design and Architecture, Kingston University, London, UK.

Routledge Research in Architecture

The *Routledge Research in Architecture* series provides the reader with the latest scholarship in the field of architecture. The series publishes research from across the globe and covers areas as diverse as architectural history and theory, technology, digital architecture, structures, materials, details, design, monographs of architects, interior design and much more. By making these studies available to the worldwide academic community, the series aims to promote quality architectural research.

Designing the British Post-War Home
Kenneth Wood, 1948–1968

Fiona Fisher

Routledge
Taylor & Francis Group

LONDON AND NEW YORK

First published 2015 by Routledge

2 Park Square, Milton Park, Abingdon, Oxfordshire OX14 4RN
711 Third Avenue, New York, NY 10017

Routledge is an imprint of the Taylor & Francis Group, an informa business

First issued in paperback 2017

British Library Cataloguing in Publication Data
A catalogue record for this book is available from the British Library

Library of Congress Cataloguing in Publication data
Fisher, Fiona.
    Designing the British post-war home : Kenneth Wood,
    1948–1968 / Fiona Fisher.
      pages   cm. – (Routledge research in architecture)
    Includes bibliographical references and index.
    1. Wood, Kenneth, 1921–Criticism and interpretation.   2. Architecture,
    Domestic–Great Britain–History–20th century.   3. Modern movement
    (Architecture)–Great Britain.   I. Title.
    NA997.W685F57 2015
    728′.37094109045–dc23
    2014040629

ISBN: 978-0-415-82354-8 (hbk)
ISBN: 978-1-138-56747-4 (pbk)

Typeset in Sabon
by Out of House Publishing

Contents

List of figures

Acknowledgements

I am grateful to acknowledge that the research for this book was completed with the support of an Early Career Fellowship, awarded by the Arts and Humanities Research Council [Grant Number: AH/I002928/1].

I would like thank the following individuals and organisations for assisting my research: Juliet Parry at Kingston University for her valuable help in preparing my grant application; Rhoddy Voremberg, Lorraine and Simon Dittrich at the Stanley Picker Trust for permission to reproduce images from the Trust's collection; Gilly Booth and her colleagues at hijack for a splendid film on Kenneth Wood and the stimulating questions that helped develop my thinking while we were making it; Tove Bellingham at Kingston Museum; Julia Rees and Jennifer Butterworth at The Kingston Society; Cara and Colin Rodger and Jen Manhire at Emmanuel Church; Christine Brown and Elaine Coulon at Barnfield Youth and Community Centre; Elaine Penn and Anna McNally at the University of Westminster Archive; Geraint Franklin at English Heritage; Robert Gallagher at British Pathé; Elspeth Millar at the British Library Sound Archive; Phil Cotterell at the Department for Transport; Mark Richards, Jack Richards and Casper Scott for their help in Saffron Walden; Michael Kurtz for kindly making his Canadian Trend House materials available to me; Elizabeth Turner at TRADA; Samantha Blake at the BBC Written Archives Centre; Gabrielle Allen at Guy's and St Thomas' Charity; Andy Moffat at The Forestry Commission; Jonathan Makepeace, Justine Sambrook and Fiona Orsini at the Royal Institute of British Architects; and Chris Thomas and Bruce Morgan for much appreciated technical support with Wood's visual archive.

My colleagues at the Modern Interiors Research Centre at Kingston University and the many researchers who have contributed to its publications and conferences have inspired me, and their work forms the foundation for my own research. I would like to thank Trevor Keeble and Emma Ferry for the interesting discussions about architecture, design and the home that set me on this path some time ago. I am especially grateful to Brenda Martin and Patricia Lara-Betancourt for their encouragement and support, and to Penny Sparke for her generosity as a colleague, which I value greatly. My research on the Picker House was furthered by that of my co-authors of

The Picker House and Collection, Fran Lloyd and Jonathan Black, whose research into Stanley Picker's art collection greatly assisted my consideration of Wood's design of the house and gallery; Rebecca Preston, whose investigations into the Picker House site and garden opened my eyes to the landscape; and David Falkner for his support of my research at the Stanley Picker Gallery.

My special thanks go to the clients, associates, and former employees of Kenneth Wood Associates who have helped me with photographs, interviews and invitations to look around their houses: Martin Warne, Jeremy Draper, Paul Willatts, Sylvia and John Disley, Elisa Money, Janet Mundy, Averil and Vincent Marks, Grainne O'Keeffe, Norman Blackburn, Alexa Woolf, Vicky Jones and Marie Davies.

I have been grateful for the support and advice of Francesca Ford, Jennifer Schmidt, Alex Hollingsworth, Louise Fox and Sadé Lee at Routledge/Taylor & Francis at different stages in the development of this book and thank Emma Hart and Chris Steel for their care with my manuscript.

Over the course of the research for this book I was privileged to spend many stimulating hours in conversation with Kenneth and Micki Wood. I am grateful for the time that they spent with me and the generous spirit in which they opened up their home and shared their professional lives.

Kenneth Wood died in January 2015, just as the preparation of this book was reaching its conclusion. I hope that it does justice to the work that he and his colleagues completed.

Abbreviations

A&B	*Architecture and Building*
A&BN	*The Architect and Building News*
AD	*Architectural Design*
AJ	*Architects' Journal*
AR	*The Architectural Review*
BCLMA	British Columbia Lumber Manufacturers' Association
BDR	*The Builder*
DPP	Mr and Mrs Disley, private papers
EMPP	Elisa Money, private papers
KWPP	Kenneth Wood, private papers
RIBA	Royal Institute of British Architects
SPPP	Stanley Picker, private papers
SPT	Stanley Picker Trust

Introduction

This book takes the work of a small, suburban architectural practice as a lens through which to explore the design of the private house in post-war Britain. The firm was founded by Kenneth Wood at East Molesey in Surrey in 1955 and evolved as a successful, non-specialist architectural and design practice. Its work included church halls, commercial offices and showrooms, schools, a youth club, a village centre, a child guidance unit, a regional headquarters for the Forestry Commission, street improvement schemes and interior design consultancy for Britain's first hypermarket group, SavaCentre. An leading area of practice in the firm's early years was the design of private houses, the majority of which were completed in Surrey's suburbs and commuter towns, between the mid 1950s and the late 1960s.[1]

Wood's firm received recognition for its work at a national level. It is, however, largely absent from British architectural histories of the post-war period. Why study the work of a small and today relatively unknown architectural firm? Although no aspect of Wood's work can be described as unique in terms of British architectural design of the 1950s and 1960s, his practice was forward-looking in its approach and displays certain characteristics that make it of wider interest and significance. Wood's desire to create flexible long-term living environments was coupled with an interest in the possibilities of new timber products and construction methods. He was among a relatively small group of British architects who began to explore and promote the structural use of timber in the design of British homes from the 1950s. His association with the Canadian lumber industries' efforts to promote timber houses to British post-war consumers helps to reveal some of the ways in which North American models of domestic design were incorporated and adapted within a British context.

Wood located his practice in suburban Surrey, just outside the major population centre of Kingston upon Thames, a town with one face turned to the country and the other toward London. Wood's firm offers an alternative vantage point to that of the city, the provincial town, or the post-war New Town from which to consider the progress of modern domestic architecture in Britain. It also presents an opportunity to re-think the suburb as a space of creative production and to explore the social and professional networks that supported modern architectural design within that arena.

Wood's professional training was completed during a period of intense debate about the basis for, and future of, modern architecture. There was, it seemed, no clear way forward, rather a series of possibilities to investigate, negotiations to be made and elements to be held in balance in continuing to develop modern architectural design to meet contemporary needs. As few opportunities had existed for modern architects to build large public buildings in Britain between the wars, private house commissions had been an important testing ground for new ideas. The period has been interpreted as one in which the architect-designed house was both widely disseminated and an indicator of broader architectural trends.[2] As F.R.S. Yorke observed in *The Modern House*,

> In so far as the modern architect is concerned the villa has had, and will continue to have, a great importance as the cheapest complete building unit for examination and experiment, and it is most often in this small structure that modern architecture goes through its complete revolution.[3]

In the years immediately after the Second World War, major public projects formed a focus for architectural progress, but as building resumed in the mid 1950s, the private house continued to provide opportunities for experimentation. For newly qualified architects such as Wood it offered a significant degree of autonomy and the chance to put into practice ideas developed during their training.

Wood's early houses were disseminated widely, in professional and popular contexts: architectural exhibitions, architecture and design journals, technical journals, home-oriented women's magazines such as *Woman*, general interest magazines such as *Country Life*, and home interest magazines such as *House and Garden*. This breadth of coverage permits an examination of the reception and mediation of Wood's work for diverse contemporary audiences. The book will consider the publishing practices and personal relationships that informed the representation of Wood's designs in different editorial environments and the intersection and shaping of architectural and domestic values within them.

Surrey proved to be a productive context for Wood's professional development, providing varied sites and a steady stream of middle-class clients for modern houses of a practical and unassuming type. How did those sites inform Wood's approach to design? Who were his clients? How did they choose between commissioning a house of 'traditional' or 'modern' design? What factors motivated them to build rather than buy their own homes? How were their tastes and aspirations reflected in the houses that Wood designed for them? And how did they go about furnishing, decorating, and inhabiting the spaces that he created? This study will address all of these questions.

The book is structured as follows. Chapter one looks at Wood's training at Regent Street Polytechnic's School of Architecture in London and his early professional experience, which included a brief but formative period working for Eric Lyons. Chapter two considers the representation of modern domestic architecture in inter-war and post-war Britain and its history in Surrey, the principal location in which Wood worked. The chapters that follow examine a selection of Surrey houses completed by Wood's firm between the late 1950s and the late 1960s. Each project is used to explore a theme: the modern house as a working environment (Whitewood); the modern house and flexible living (Wildwood); the modern timber house (Design 109); the exhibited house (Vincent House); the developing house (Fenwycks); the show house (Hampton House); the converted house (Torrent House); and the house for art (Picker House). Collectively, these projects map the development of Wood's practice in relation to some of the wider themes and concerns of modern architecture in Britain at the time.

Notes

1 Most of these houses were completed in pre-1965 Surrey. In 1965 the London Government Act of 1963 came into force and Surrey was reduced in size with the creation of new London boroughs: Kingston, Merton, Richmond and Sutton. Croydon was also enlarged at that time.
2 Alan Powers, *The Twentieth Century House in Britain: From the Archives of Country Life* (London: Aurum Press, 2004), 13; Beatriz Colomina, 'The media house', *Assemblage* 27, Tulane Papers: The Politics of Contemporary Architectural Discourse (August 1995): 55–66.
3 F.R.S. Yorke, *The Modern House*, rev. edn. (London: Architectural Press, 1951), 5.

1 Kenneth Wood
An introduction

Professional training

Kenneth Brian Wood was born into a working-class family at Silvertown, East London in 1921 and spent his childhood in London and Kent. Following a secondary education at Dartford Grammar School he continued his studies at The Regent Street Polytechnic's School of Architecture in London. He initially enrolled in the evening school to train in engineering services (1937–39) and worked by day as a junior draughtsman for Matthew Hall and Company, a large engineering firm based in London's Dorset Square.[1]

Wood's studies were interrupted by the outbreak of the Second World War, during which he trained in the Royal Air Force as an aircrew fighter navigator and in airfield control. Between January 1944 and April 1945 he was seconded to the Air Transport Auxiliary where he served as a pilot in the ferry pools at Thame, White Waltham and Hawarden.[2] It was during his military service that he began to develop a broader interest in design, sketching out ideas for 'minor things' as a relaxing way of occupying his mind in free moments.[3] In 1944 he married Betty 'Micki' Sergeant at Dartford in Kent and the two embarked on a successful lifelong partnership. Micki was for many years the backbone of Wood's architectural practice; the safe and accomplished pair of hands that kept the office running smoothly and maintained a careful eye on its financial stability.[4]

After the war Wood resumed his working life at Matthew Hall and returned to his engineering training with greater maturity and a new found sense of confidence, responsibility and ambition.[5] It was at that time that he began to pursue his interest in design more seriously, entering and winning a Central Institute of Art and Design prize in 1946 for his practical reconsideration of an everyday household object – a meat carving dish. In the same year a brief secondment to Blomfield's architectural practice further expanded his professional horizons.[6] His experience working with student architects at the firm, some of whom were also training alongside him at Regent Street, convinced him to change professional direction. In 1948, on completing his engineering qualification, Wood reapplied to the Regent Street Polytechnic and was accepted to study architecture, commencing in the evening school as one of three older ex-servicemen on the course in the autumn of that year.[7]

Although the direct influence of what has come to be called the Modern Movement in architecture was fairly limited in terms of what was built in Britain in the inter-war years, its influence in Britain's architecture schools was significant.[8] Trevor Dannatt, who trained at the same school of architecture from 1938, recalls that the course had a practical emphasis that was supplemented by lectures on the theory and literature of architecture.[9] He remembers reading Le Corbusier's *Towards a New Architecture* during his first year of study and of being conscious of the existence of a new movement in architecture at that time.[10]

On Wood's recommended reading list, as he began his architectural training at Regent Street ten years later, were a number of introductory and historical texts. Apart from Le Corbusier's *Towards a New Architecture*, there were Frederick Gibberd's *The Architecture of England: from Norman Times to the Present Day*, Nikolaus Pevsner's *European Architecture* and J.M. Richards' *Introduction to Modern Architecture*, along with technical works and studies of allied subjects, including fine art and graphic design. Contemporaries of Wood who attended Regent Street in the late 1940s remember it as a conventional establishment with a practical focus on modern design and construction.[11] Students were encouraged to explore the development of modern architecture and were expected to design in the contemporary idiom, creating buildings that fulfilled the social requirements of the day through the employment of rational and scientific principles.

The curriculum for Wood's first three years of study included: studio design, construction, theory of structures, theory of architecture, colour, museum studies and outdoor sketching, materials, land surveying, measured drawings, building regulations and theory of planning. In years four and five he went on to study acoustics, sanitation, specifications, heating, lighting, ventilation and professional practice, some of which he was already familiar with through his earlier training. A component of the course that Wood particularly valued and enjoyed was the historical study of pre-Gothic, Gothic and Renaissance architecture.

As modern architecture evolved in plural forms after the Second World War, the pressing question for young architects such as Wood was not whether to become a modern architect but what type of modern architect to become. In March 1950, in a lengthy article in *The Architectural Review*, J.M. Richards examined some of the most significant local, national and international architectural tendencies that had emerged in the previous decade and questioned whether common principles or directions might be identified to help guide future development.[12] He was chiefly concerned with the 'premature stylization' of modern architecture and identified four progressive architectural strands: a 'mechanistic' strand that acknowledged 'increasing mechanization and the industrialisation of building' as the basis for future architectural development; a 'post-cubist' strand, evolving from 'the pure abstraction of the thirties' within which he included the British work of Tecton; a 'regional organic' strand that attempted to 'substitute regional character for the international style, as a way of building on existing traditions without abandoning modern freedom of planning and technique'; and an 'empirical organic' strand that included recent developments in Swedish architecture.[13]

Swedish architecture had been examined by the journal at some length three years previously, initially with reference to modern houses designed by Sven Markelius, Stüre Frölén and Ralph Erskine, which were taken to exemplify efforts 'to humanise the aesthetic expression of functionalism' through a reconsideration of human habits and emotional needs.[14] A second article on what the journal termed 'The New Empiricism' followed in January 1948 and effectively proclaimed the first wave of modern architecture over: 'The first excitement of structural experiment has gone and there is a return to workaday common sense. There is a feeling that buildings are made for the sake of human beings rather than for the cold logic of theory.'[15] This new Swedish tendency, it argued, was expressed in freer planning, a more diverse approach to fenestration, widespread use of indigenous materials inside and out, attention to the relationship between the site and the surrounding landscape, consideration of external landscaping as part of the overall design, an emphasis on the creation of atmosphere, reflected in greater 'cosiness' in domestic work and, among 'the more sophisticated', the mixing of furniture of different styles.[16] This Swedish model – with its balancing of visual and functional, social and cultural priorities – had particular resonance for Wood as he began to develop his early domestic work and can be distinguished from the picturesque approach that Reyner Banham later described in *The New Brutalism* (1966) in relation to the style of Swedish-influenced housing that emerged in Britain's New Towns.[17]

Eric Lyons and Span

During his time at Regent Street, Wood worked as an assistant to Dudley Marsh, who was based at Herne Bay in Kent, with Farmer and Dark – where he was involved in designs for a power station at Fahahil in Kuwait – and in the architects' department of the recently formed Regent Oil Company at Park Street in Central London. On completing his formal training his first position was with Eric Lyons at his practice at East Molesey in Surrey, where he worked from around 1953 to 1956. Although only nine years older than Wood, Lyons was from a different architectural generation. A graduate of Regent Street Polytechnic in the early 1930s, he had gone on to work with Maxwell Fry and Walter Gropius, and with Andrew Mather, before joining Geoffrey Townsend to design speculative houses just before the war.[18] Wood shared many of Lyons' values and having identified him as a prospective employer made a speculative enquiry to see whether there might

Figure 1.1 Eric Alfred Lyons, Parkleys, Ham Common, 1956.
Photographer: Sam Lambert. Credit: RIBA Library Photographs Collection.

be a suitable opening for him. Contacting Lyons proved to be a significant decision, placing him at the firm during an important period in its development of a new vision for modern suburban housing.[19]

Wood's time with Lyons coincided with the firm's work on the Parkleys housing development at Ham, near Richmond; a speculative scheme of flats and a small parade of shops on an eight acre landscaped site and one of a number of modern housing estates completed by Span in the late 1950s (see Figure 1.1). The originality of the scheme was soon recognised and emulated by other house-building firms operating in and around London, among them George Wimpey. Writing in 1959, Trevor Dannatt noted the careful attention that had been given to the layout of Parkleys and appreciated the way in which the relationship between the buildings and landscape helped to create a sense of identity for each element within the overall scheme: 'Each courtyard has been given a slightly different character by its landscaping. Links have been made between courtyards by opening up the ground floor of some of the blocks.'[20] Wood was sympathetic to Lyons' respect for local conditions and architectural traditions – values that not only shaped the landscaping at Parkleys, but were also displayed in the tile-hanging, which Ian Nairn described as 'a genuinely twentieth century comment on the native Surrey style'.[21] While working for Lyons Wood was also involved

Figure 1.2 Kenneth Wood at home at Merton House, East Molesey in the 1960s.
Credit: Courtesy of Kenneth Wood.

in a number of other projects, among them housing for the Soviet Trade Delegation at Millfield Lane, Highgate (1957) and the detailing of Lyons' early designs for a house for his business partner Geoffrey Townsend.[22]

Wood was elected an Associate of the Royal Institute of British Architects (RIBA) in February 1954, just as rationing and restrictions on private building were coming to an end. Having worked throughout his engineering and architectural training, he had good practical experience and knowledge of building to draw upon. Keen to get on and seeing the potential of the local area, the following year he set up in private practice close to Lyons at East Molesey (see Figure 1.2). Lyons continued to employ him on a part-time basis until he was sufficiently established to concentrate on his own practice full time. Their professional paths crossed again in the 1960s, when Lyons invited Wood to join the panel of architects for the Redhill Wood area of Span's ambitious New Ash Green scheme, a modern village on a 430 acre site in Kent. His fellow architectural panellists were Richard and Su Rogers, Roy Stout and Patrick Litchfield, Walter Greaves, Peter Aldington, and Carol Møller.[23] Redhill Wood was planned as a low density area of fifty architect-designed houses, but did not go ahead in its intended form.[24]

Townscape

In his first few years in practice Wood worked primarily in Surrey's suburbs and commuter towns. The writings of Ian Nairn and Gordon Cullen were strong reference points for his work during that period. From the late 1940s, Cullen and others had begun to examine the quality of everyday spaces. Through their discussions of the degraded character of many of Britain's postwar environments the 'townscape' concept evolved.[25] Its supporters promoted a visual approach to design and planning that valued the historic fabric of local environments.[26] Their critique gained momentum just as Wood began his professional life. In June 1955, under the authorship of Ian Nairn, *The Architectural Review* published 'Outrage', a special issue on the subject of town planning that was followed, in 1956, by a second special issue, 'Counterattack against Subtopia'. Nairn's call for 'a visual conscience to partner the social conscience that has grown up in the last century' struck a chord with Wood, who had both issues of the journal specially bound.[27] In 1957 the Civic Trust was formed to challenge 'the characterless urban sprawl which is disfiguring both town and country'.[28] Headed by Misha Black, a street improvement scheme at Magdalen Street in Norwich (1959) was its first attempt to create a strong contemporary visual identity for a local shopping area.[29] The idea was soon taken up elsewhere and Wood became involved as coordinating architect for a scheme to improve his local shopping street, Bridge Road at East Molesey, and in a similar project at Chertsey.[30] The Chertsey scheme was intended to bring immediate benefits to the town centre in advance of a major development programme, for which Eric Lyons was the consultant architect.[31] Following principles established by the Civic Trust at Magdalen Street, Wood aimed to rationalise town centre signage and to introduce a more unified appearance to its buildings through the use of a limited palette of exterior colour and graphic treatments. His proposal document for the scheme, which focused on drawing out the architectural qualities of each building, reflects his commitment to good manners in urban design. In it he wrote: 'In a busy, lively shopping area the buildings generally should be a background for the activity taking place and not jostle assertively for individual attention; this is the proper province of window display.'[32] His care for the environments in which he worked is reflected in a number of awards and commendations made to his firm under the Civic Trust Amenity Awards scheme. Launched in February 1959, the scheme aimed to encourage local work 'which can be said to make a desirable contribution to the civic and rural scene as a whole' and to promote an interest in 'all other types of physical development making a desirable contribution to the civic and rural scene as a whole'.[33]

Suburban churches: 1950s

The demand for new social buildings to meet the needs of Surrey's rapidly expanding suburban communities presented Wood with some of his earliest opportunities as an independent architect in private practice. Churches, in particular, were a productive source of early work. Between 1931 and 1933 the population of the Church of England parish of Christ Church at Surbiton expanded by 6,000 people and 1,700 houses. Christ Church and its neighbouring parish of St Matthew's responded by building two new churches, St George's at Hamilton Avenue, Tolworth (1934) and Emmanuel Church at Grand Avenue, Tolworth (1935). The speed of population growth was such that within a few years of completion they were both too small to meet the needs of local parishioners, a pattern of development that can be seen in other suburban locations. Wood was engaged to extend both churches in the 1950s.

In the absence of suitable public buildings, suburban churches often fulfilled a variety of local needs. Many not only functioned as places of worship and as social or recreational spaces but also housed community services such as healthcare provision. Edward Mills, whose influential book *The Modern Church* (1956) was reprinted three times in the 1950s, emphasised this broader role in his discussion of contemporary church design.[34] Through his church commissions Wood developed an interest in ecclesiastical architecture and became involved in the architects' group of the Christian Teamwork Trust, participating in a conference on the subject of 'The Architect and the Church' in 1960 and in an exhibition of church design in 1964, for which Mills was on the hanging committee.[35] Wood's early church projects reveal an interest in modern timber construction methods and a concern with design for expansion that he pursued in his later domestic work.

Published sources on Wood's additions to St George's and Emmanuel are helpful in situating his practice within the context of national architectural developments, and are also indicative of contemporary attitudes to the type of suburban locations in which he was working at that time. The discouraging context for his new church hall for St George's was described in the *Architects' Journal*:

> The site is surrounded by a very ordinary building estate [of] semi-detached houses and the church, built in the 1930s, is equally ordinary; in style more Jacobean than anything else, in character altogether too fussy and pretentious for its diminutive scale and in material a rather unattractive dark red brick.[36]

In 1950 the church committee had considered replacing the church, but when that had proved cost-prohibitive, they decided to build a new church hall instead. Wood was subsequently commissioned and designed a low-cost building, in load-bearing brick, with a highly glazed facade. The hall was erected to the east of the church, at right-angles to it, with a lower height lobby connecting the two buildings. The hall was designed for expansion at a future date and Wood planned the interior to allow the reuse of materials to create aesthetic continuity between the old and new elements. Plans for the hall were published in *The Architectural Review* in January 1956.[37] In 1957, the finished building featured in the *Architects' Journal* in a 'criticism' piece by J.M. Richards, with a response by Wood in the following edition. A further article on the reception of the building was published in the *Architects' Journal* in January 1958.[38]

Richards commented on the lack of architectural merit of the original church, noting that Wood had 'quite rightly let his new building make its own contemporary statement. But, also quite rightly,' he added, 'not in too loud a voice'.[39] The structure, he felt, was 'reasonable and economical' and he appreciated 'the choice of materials to reduce maintenance costs' but took issue with the front elevation, criticising 'the false architectural character given by the use of glass as a cliché, not as something arising from the needs of the building' and inviting Wood to justify its inclusion on functional grounds.[40] Wood's spirited reply offers the first documented evidence of his professional views at the time.

> Buildings are [he responded] or should be designed for a purpose and they should come to life fully when occupied, not be designed as precious set-pieces where usage is deplored. It is on this level, too, that they must be assessed as well as that of pure art, to see how they measure up as *architecture*.[41]

In response to Richards' main point of criticism, regarding the glazed facade and the lack of privacy that it afforded to the interior, Wood replied that the area to the front of the new building was private and had yet to be planted. He also drew attention to his use of clerestory lighting to the rear, to render the facade less transparent. He maintained his right to an artistic basis for his work, arguing that the architect 'must preserve his right to make design decisions that are not based purely on function, but a proper balance of both function and intuitive sensibility' – a position that he maintained throughout his career.[42]

Although the design was initially controversial, the hall came to be valued for its low running costs and efficient use of space.[43] Other than a few minor complaints, including some grumbling over the quality of the PVC floor for dancing, responses from users appear to have been favourable. The vicar approved of the inviting appearance of the building when it was illuminated for social events at night and a doctor, who held a clinic there, commented on its 'air of well-being' and the positive affect that this was having on patients.[44]

The conditions for the development of Emmanuel Church were fairly similar. The church, designed by J. Rycroft of Henry Boot and Sons, was consecrated in 1935.[45] By 1941 it was too small to meet local needs and the church committee began to investigate the possibility of building on a new site, or redeveloping the church and its neighbouring timber hall.[46] Initial plans failed and a second attempt was made in 1956, when Wood was

Figure 1.3 Kenneth Wood, Emmanuel Church, Tolworth. View of the timber frame
     and screen during construction, *c.* 1958.
Credit: Courtesy of Kenneth Wood.

employed on the basis of his preliminary drawings and imaginative design.[47]
The eventual contract cost was £13,500 and included the church furniture,
which Wood designed. The church was re-dedicated in 1958.[48]

Wood's plans for Emmanuel involved demolishing the small timber hall
and altering the liturgical orientation of the church to allow the creation of
an entrance lobby to link it to a new church hall. To connect and visually
integrate the old and new elements, the original church building was covered with a large triangular screen, in cedar and Douglas fir. The design was
practical, economically viable and visually striking, giving the church greater
prominence within its local setting. From the exterior, the impression given
was that of a modern A-frame church (see Figure 1.3), a form favoured by
architects of churches in North America's mid twentieth-century suburbs.[49]
There was, however, no pretence to Wood's architectural solution: the timber screen was louvred, leaving the original 1930s building visible beneath
the new frame and its roof was only partially tiled to allow light to enter the
baptistery below. The relationship between old and new was equally explicit
within the interior, where the timber construction of the new hall, entrance
lobby, and sanctuary was visible throughout.

As was the case with St George's, articles on the project emphasised the
poor appearance of the original church. *Building Materials* described it as

'ungainly and of little character' and *Interbuild*, which welcomed 'the functional honesty' of Wood's timber screen, similarly deplored its 'lamentable ugliness'.[50] *Architects' Journal* focused on the unfortunate circumstances that had made such alterations necessary:

> All too many buildings in Britain are structurally satisfactory but visually appalling, and financial resources are not available to enable them to be re-built. One such example was Emmanuel Church, Tolworth, Surrey. Fortunately, in this case, funds were available to build alongside a church hall, and the architect, Kenneth Wood, insisted in using this as an opportunity for drawing a veil – in cedar – over the main facade ... The result, however false, is more attractive than the original.[51]

Both churches are referenced in the London South edition of *The Buildings of England*, which recognised their contribution to the local environment:

> Improvement has come from an unexpected direction, in additions to two churches by Kenneth Wood, 1957–59. Both are poor wee things of the 1930s; now St George has a handsomely detailed flat-roofed church hall attached to it, and Emmanuel has been given a very successful false w. front.[52]

Descriptions of the two projects reflect the mood of contemporary discussions about the visual quality of the type of suburban environments in which Wood was working. In the case of the 'false' character of his 'veil' over Emmanuel Church they also suggest that some anxiety was felt by architectural critics when confronted by a successful building that betrayed the slightest hint of lack of integrity as a complete contemporary architectural statement.

These early projects brought Wood regular church inspection and were no doubt also helpful when it came to winning further commissions for social buildings. Among those completed Wood in the 1950s and 1960s were two parish halls for St Mary-at-Finchley, north London (1958), a village centre for Oxshott (1962), and Barnfield Youth Club (1969) at Kingston, which he designed in association with J.H. Lomas, the borough architect (see Figure 1.4). The latter, located in a residential area of north Kingston, was designed for expansion. The site on which it was built had been previously used for prefabricated housing and the ground conditions were poor, with a high risk of settlement. The building was conceived as a lightweight, flexible timber-framed structure, on a raft type foundation.[53] In 1970 the building was commended under the Civic Trust awards scheme:

> This is a good example of a low-budget building achieving a high standard of design and giving a lead in a rather neglected area ... There is an approriately rugged simplicity throughout this design, which should

Figure 1.4 Kenneth Wood in association with J.H. Lomas, Barnfield Youth Club, Kingston.
Credit: RIBA Library Photographs Collection.

resist the expected wear and tear. The materials are economical and the form is very expressive of its purpose.[54]

Wood exhibited the project at the Royal Academy Summer Exhibition in the same year.[55]

Private houses

Private houses represent the other main strand of work completed by Wood in the 1950s and 1960s and provide the substance of this study. From 1 April 1954 any individual could obtain a licence from the Ministry of Works to build a house of up to 1,500 square feet. In his first five years in private practice Wood designed five Surrey houses – Whitewood at Strawberry Hill (1958), Wildwood at Oxshott (1958), Vincent House at Kingston (1959), Nathan House at Oxshott (1960), and Fenwycks at Camberley (1960). These early houses were followed by Hampton House at Hampton (1961), Atrium at Hampton (1962) (Figure 1.5), Tanglewood at Hampton Hill (1962), Oriel House at Haslemere (1963), Dykes at Virginia Water (1963)

Figure 1.5 Kenneth Wood, Atrium, Hampton, 1962.
Credit: Courtesy of Kenneth Wood.

(Figure 1.6), Torrent House at Hampton Court (1964), Little Boredom at Cobham (1968), and Picker House at Kingston (1968).[56] In 1961, an important year for the firm, Vincent House was selected for display at Architecture Today, a highly selective exhibition of contemporary British architecture that was timed to coincide with the Congress of the International Union of Architects (CIAM), which was being held in London in that year. In the same year, Hampton House was published as the popular magazine *Ideal Home*'s House of the Year.[57]

When viewed collectively, Wood's clients appear forward-thinking in their domestic aspirations and tastes. Several were professionals working in the creative fields of art, design or music; among them: a professional flautist, a former ballerina, three employees (aircraft designers and engineers) of the Hawker Aircraft Company at Kingston, a graphic and packaging designer, a fine artist, an interior designer and ceramicist, a music arranger, a professional singer, and the conductor of the BBC Welsh Symphony Orchestra. Some came to Wood through personal recommendation, others through enquiries made at the Royal Institute of British Architects, and one through a chance conversation in a Surrey council office queue. Their budgets

Figure 1.6 Kenneth Wood. Dykes, Virginia Water: a side elevation, 1964.
Credit: RIBA Library Photographs Collection.

varied enormously. Wood's first private house commission, Whitewood at Strawberry Hill (1958/1,300 square feet), for his friends Peter and Vicky Jones, was completed for under £3,200.[58] In contrast, the Picker House at Kingston (1968/4,975 square feet), for businessman and art collector Stanley Picker, cost close to £81,000.[59]

Social and professional networks have been identified as a significant factor in the patronage of modern architectural design in the inter-war period; a period in which clients for modern houses were not always motivated by the desire to commission a building of modern design, but often by an aspiration to live in a less formal and restrictive manner than that of contemporary middle-class convention.[60] Although Wood's clients of the 1950s and 1960s share certain characteristics with those early patrons of modern architecture – a number, for example, moved in similar social circles as a result of their professional interests – there is greater evidence among them of a commitment to modern design. That commitment was in certain cases related to their professional training in art or design, and in others appears more generally indicative of the growing appeal of modern architecture to a post-war middle class.

Other motivations for clients to commission their own homes were the specific social and functional needs that they had for them. In a number of instances those requirements could not be satisfied by the existing housing stock and demanded a more individual approach to design. Several of Wood's clients wanted to work at home and the spaces that he created for them responded to their practical requirements as well as the different attitudes that they held toward the place of work within their everyday lives. Acoustics were a significant design consideration in the case of two houses that Wood designed for musicians. Stephan Quehen, his client for Tanglewood, was a music arranger and needed a suitable space in which to work without disturbing his family and neighbours. Wood planned his single-storey house with buffer zones around the study to minimise the transfer of sound within the interior. The acoustic performance of the interior was also a concern for professional flautist, Douglas Whittaker, Wood's client for Vincent House, for whom he designed an open-plan living area with a music dais for rehearsal and performance.

Personal hobbies and interests, some of which had a semi-professional status, also informed the development of certain designs. For Whittaker, a serious photographer, Wood designed a dark room and workshop. Nigel and Elisa Money were keen ornithologists and Wildwood, the galleried house that Wood created for them, was designed to give good views of the surrounding trees from all parts of the interior. Stanley Picker's desire to create an informal domestic environment in which to enjoy his collection of contemporary and modern painting and sculpture was at the heart of his brief for the luxury house at Kingston, which Wood completed for him in the late 1960s, and the private art gallery, which he added to the grounds in the 1970s.

Different ideals of taste and comfort shaped Wood's clients' visions for their new homes. While some self-consciously aimed to create a context in which to live in a new way, others were concerned to maintain continuities of domestic practice within a more efficient and modern architectural setting. Similarly, whereas some intended to furnish their new homes with furniture of modern design, several preferred to mix older items with new acquisitions. The Whittakers, for example, briefed Wood to design a house in which their antique furniture and paintings would look and feel at home. In contrast, almost everything for Stanley Picker's house at Kingston Hill, from furniture to tableware, was either commissioned or specially chosen for the interior.

Despite the advances made by modern architecture within the sphere of post-war public sector design, completing houses and housing in the modern style remained a challenge for architects in private practice. It required fortitude, creativity, and, according to Eric Lyons, led him to develop 'a certain cunning; how to fake drawings, provide ambivalent information, accept irrelevant compromises'.[61] Wood became equally practised at circumventing potential planning problems and negotiating the additional difficulties involved in building on Surrey's private estates, where residents had a strong voice in the planning process. Wood's personal recollection of his early years in practice is that around half of his designs for private houses went to appeal on the basis of concerns about their suitability for the local environment.[62] Peter Womersley, who operated on a similar scale to Wood, and was also designing with timber in the 1950s, faced similar problems:

> If your practice is a new one based for a start on private houses, you are particularly vulnerable, since there still appears to be an antipathy towards the modern house which does not apply to other building forms. The number of houses designed by the firm and rejected almost equals the number put up, and the reason for the abandoned project is always the same – 'unsuitable for the environment'.[63]

Womersley also considered the nature of private work, commenting:

> Modern house design could, I think, be called a specialised branch of architecture – it is largely the province of the younger architect starting a practice; it entails extraordinary involvement with a personal client; it falls foul of many planning officers and building societies; and the demands made upon a practice, as far as time, drawings and supervision are concerned, are still totally incommensurate with the financial return.[64]

Wood's original architectural intentions were inevitably compromised on occasion. Vincent House, which was built on a private estate at Kingston, is a case in point. Wood side-stepped a potential problem with the split pitched roofline, presenting a decidedly ambiguous drawing to the local residents' association, in which its unconventional form was partially obscured by clouds. However, he was unsuccessful with the local authority in the case of his structural plans; he had intended to build in timber but concerns over fire resistance led to his eventual use of concrete.

Wood's interest in architectural history and vernacular traditions informed his work. It is telling that of all the student projects that he completed during his training, a comparative study of the Renaissance architecture of Italy and France is one of the few items that he retained into his retirement years (see Figure 1.7). A challenge that preoccupied him throughout his career was that of how to create flexible buildings. In the case of the individual dwelling, he was concerned both with the flexibility of the plan and the potential for economic expansion to create houses fit for use at different stages of family life – a modern equivalent to the accretive domestic architecture of the Surrey hall house, with which he was familiar. His approach to materials is also closely related to his desire to create adaptable houses, suitable for long-term use. Several of his designs combined

Figure 1.7 Kenneth Wood. Student drawing of Francesco Borromini's church of San Carlo alle Quattro Fontane, Rome. From a joint project of 1951 with Donald Cheney and David Hinge.

Credit: Courtesy of Kenneth Wood.

structural brick and timber. In his design for Fenwycks the timber elements of the building were positioned to allow cost-effective expansion to a phase two plan. Wood's use of materials for the interiors of his houses was similarly informed by a desire to create spaces with enduring values that would age and evolve in conjunction with their occupants. This is evident in his preference for natural materials such as wood, cork, brick and slate and his more cautious approach to the use of modern materials with fashionable appeal, which is something that distinguishes his work out from that of

Figure 1.8 Kenneth Wood. Extension to Wychbury Cottage, St George's Hill, viewed
        from the garden.
Credit: RIBA Library Photographs Collection.

architects working in the 'contemporary style' that developed in Britain in
the 1950s.

Not all of Wood's clients had the resources or the inclination to build or
furnish their homes from scratch and a number of projects involved extending or reconfiguring the interiors of older properties to accommodate the
more informal middle-class lifestyles that evolved after the Second World
War. Notable among the firm's projects of this type was Wychbury Cottage
at St George's Hill, Weybridge (1967), for John Fozard, a leading figure in
aircraft design.[65] Wood replanned and substantially enlarged his small villa,
adding a suite of garden rooms in a highly glazed addition that several years
after its completion was described as 'uncompromisingly modern in its lines'
(see Figure 1.8).[66]

Later projects

By the early 1960s Wood had begun to style himself as an architect and
design consultant, reflecting the nature of his firm's work at that time and
his intention to maintain its breadth of practice within an expanded design

Figure 1.9 Kenneth Wood. St Paul's School, Kingston.
Credit: Colin Westwood/RIBA Library Photographs Collection.

field in which interior design, landscape design and design consultancy had emerged as significant areas of professional activity. The direction of Wood's practice changed in that decade as it expanded and moved into local authority work. That expansion came at a challenging time for small architectural firms. In February 1970, *Design* magazine published a piece entitled 'Down Go Architects Commissions' in which it commented on figures issued by the Royal Institute of British Architects, which showed that the value of new commissions for private sector offices had dropped by 17 per cent between the second and third quarters of 1969.[67] Although Wood's firm had work in development, the phone began ringing with cancellations.[68] Projects that were already under way were completed, among them a new school, St Paul's Church of England Junior School at Kingston (1972), which was designed as a cluster of twenty-one hexagonal classrooms, grouped around courtyard teaching areas and a central assembly hall (see Figure 1.9). The building was commended by the Civic Trust in 1973 and the citation acknowledged its sensitive design and planning, describing it as

> a gentle building which settles comfortably on its site amongst the surrounding houses like a piece in a jig-saw puzzle. Its scale is truly domestic and must induce reassurance in the most timid child. It has

been designed with the greatest respect for both its occupants and surroundings.[69]

In the mid 1970s Wood's practice moved into new territory, designing the interiors of several of Britain's first hypermarkets, for SavaCentre Limited, the jointly owned company that was established in 1975 by the high street retailer British Home Stores and the supermarket chain J. Sainsbury. As design consultant to the firm, Wood completed the interiors of the first SavaCentre store at Washington New Town in Tyne and Wear, which opened in November 1977. Built on an exceptional scale, the store had parking for over a thousand cars and a sales floor of 6,300 square metres, with specialist counters and a self-service cafeteria. Further stores followed and the firm completed the interior design of those at Hempstead (1978), Basildon (1980) and Calcot (1981). As Wood's practice grew, he ceased trading under his own name and developed his firm under the name Kenneth Wood Associates. At its peak the practice employed six assistant architects, an office manager and a copy typist. Although Wood can claim principal authorship of the houses discussed here, those who worked with him – Gordon Youett, Tony Jeerkens, Martin Warne, Derek Burton, Ivan Hollow, Peter Greiner, Jeremy Draper, Roger Jones and Wood's son, Nigel – had an important hand in the development of the firm's work and identity.[70]

Wood gave his architects and student architects the space and support to pursue their own ideas. Jeremy Draper, who trained in architecture at Kingston College of Art and worked for Wood in the late 1960s, has recalled that he was attracted to the firm because he was after 'something more imaginative' than the type of work that he was doing at Barber, Bundy and Greenfield at Dorking. He remembers his years at Wood's firm with affection, describing a family-like atmosphere, with plenty of larking about – cartoon lampoons being a speciality – but also clear leadership, focus and ambition from Wood, who was 'meticulous', 'detailed to the last screw' and 'scathing if you were not up to the mark'.[71] It is this attention to detail and willingness to go to great lengths to satisfy private clients that distinguishes the work of Wood's firm in an area of the market in which it was notoriously difficult to make a profit. The same level of attention can be seen in the firm's commercial work and is reflected in its strength in designing high-quality, bespoke social buildings.

Wood ran Kenneth Wood Associates until 1984, when he established a partnership with local architectural firm, Spicer Kapica, under the name Spicer Kapica and Kenneth Wood Associates, from which he retired in 1994. Wood died in January 2015, just as this book was nearing completion.

Notes

1 Wood's interest in engineering was perhaps inspired by his father, who was a motor engineer and pursued a wider interest in design on a semi-professional basis.

2 Phil Cotterell, Aviation Regulatory and Consumer Division, Department for Transport, email correspondence, 5 August 2009.
3 Kenneth Wood, in discussion with the author, March 2012.
4 Although much of her working life was spent managing the architectural practice, Micki pursued her own artistic interests on a semi-professional basis and encouraged her brother, John Sergeant (1937–2010), to take up art professionally.
5 Kenneth Wood, in discussion with the author, October 2008.
6 Wood has recalled only that the firm to which he was loaned as a draughtsman was 'Blomfield's'. It seems likely that it was the firm of Reginald Blomfield, which was continued by Austin Blomfield after his father's death in 1942. Kenneth Wood, in discussion with the author, October 2008.
7 John Walkden was Head of School for the duration of Wood's study. Gordon Toplis ran the day school and Robert Meadows was in charge of the evening school. The Polytechnic's School of Architecture was recognised by the Royal Institute of British Architects. Completion of the five-year Diploma in Architecture exempted students from the final RIBA exam. After passing the Diploma and completing a further examination after a year in professional practice graduates were entitled to be added to the statutory Register of Practising Architects.
8 On the influence of the Modern Movement see: Alan Powers, *Modern: The Modern Movement in Britain* (London and New York: Merrell, 2007); Susannah Charlton, Elain Harwood and Alan Powers, *British Modern: Architecture and Design in the 1930s* (London: Twentieth Century Society, 2007); David Dean, *The Thirties: Recalling the English Architectural Scene* (London: Trefoil Books, in association with RIBA Drawings Collection, 1983); James Peto and Donna Loveday, eds. *Modern Britain, 1929–39* (London: Design Museum, 1999); Elizabeth Darling, *Re-Forming Britain: Narratives of Modernity Before Reconstruction* (London: Routledge, 2007). On the influence of modern architecture in British architectural schools in the 1920s and 1930s see, Mark Crinson and Jules Lubbock, *Architecture. Art or Profession?* (Manchester and New York: Manchester University Press/The Prince of Wales Institute of Architecture, 1994), 100–8.
9 Trevor Dannatt, interview by Alan Powers, 2001, tape F11643, Architects' Lives Series, British Library.
10 Trevor Dannatt, interview by Alan Powers, 2001, tape F11643, Architects' Lives Series, British Library.
11 Max Neufeld (student at the Regent Street Polytechnic's School of Architecture from 1949) and Martin Frishman (student at the Regent Street Polytechnic's School of Architecture from 1954), interviews by Anna McNally, 2010 and 2011, University of Westminster Archives; Trevor Dannatt, interview by Alan Powers, 2001, tape F11643, Architects' Lives Series, British Library.
12 J.M. Richards, 'The next step?', AR (March 1950): 165–81.
13 Richards, 'The next step?', 169–78.
14 'The new empiricism: Sweden's latest style', AR (June 1947): 199–204, on 199.
15 'The new empiricism', AR (January 1948): 9–22, on 9.
16 'The new empiricism', 9–10.
17 Reyner Banham, *The New Brutalism, Ethic or Aesthetic* (London: Architectural Press, 1966).
18 On Lyons' training and early career see Neil Bingham, 'The architect in society: Eric Lyons, his circle and his values', in *Eric Lyons and Span*, ed. Barbara Simms (London: RIBA, 2006), 1–21.
19 On Span and suburban living see Alan Powers, 'Models for suburban living', in *Eric Lyons and Span*, ed. Barbara Simms (London: RIBA Publishing, 2006), 23–33.

20 Trevor Dannatt, *Modern Architecture in Britain* (London: BT Batsford, 1959), 160.
21 Ian Nairn and Nikolaus Pevsner, eds. *The Buildings of England: Surrey*, revised by Bridget Cherry (New Haven and London: Yale University Press, 1971), 75.
22 Kenneth Wood, in discussion with the author, October 2008.
23 Richard and Su Rogers had been working in private practice, mainly housing, since 1962, as had Roy Stout and Patrick Litchfield. Greaves trained at Regent Street Polytechnic's School of Architecture and went on to work for the London County Council (LCC) before going into private practice, specialising in residential work. Peter Aldington had worked for the LCC and TRADA (the Timber Research and Development Association) before establishing his practice in 1963 and had completed three houses at Haddenham. Carol Møller trained at the Architectural Association and specialised in landscape design, working for Brenda Colvin and Sylvia Crowe before marrying Aage Møller and establishing a private practice. Redhill Wood Leaflet, KWPP.
24 On New Ash Green see, Patrick Ellard, 'New Ash Green: Span's "latter 20th century village in Kent"', in *Eric Lyons and Span*, ed. Barbara Simms (London: RIBA, 2006), 73–4.
25 Cullen's book, *Townscape*, published in 1961, drew on his earlier writings in *The Architectural Review*. A special issue of the *Journal of Architecture*, Townscape Revisited, reconsidered the townscape movement in 2012. See, *Journal of Architecture* 17/5 (2012).
26 Matthew Aitchison, 'Townscape: scope, scale and extent', *Journal of Architecture* 17/5 (2012): 621–42.
27 Ian Nairn, 'Ian Nairn', *AJ* (19 January 1956): 87.
28 Duncan Sandys quoted in Joe Moran, '"Subtopias of good intentions": everyday landscapes in postwar Britain', *Cultural and Social History* 4/3 (September 2007): 401–21, on 411.
29 A short promotional film, *The Story of Magdalen Street*, was made by the Civic Trust. East Anglia Film Archive, cat. 304, University of East Anglia, http://www.eafa.org.uk/catalogue/304, accessed 17 August 2013.
30 Other architects in Wood's circle became involved in nearby schemes, including Alan and Sylvia Blanc and Gerald Davis, who completed a Wimbledon Improvement Scheme in 1964.
31 Plans included the creation of a Town Hall Square to give a civic focus to the town, with rerouted traffic and the formation of new shopping and housing zones in which pedestrians and vehicles were to be separated.
32 'Chertsey Street Improvement Scheme', 3. KWPP.
33 The awards were initially made triennially. The first awards covered buildings completed between May 1956 and May 1959. *AJ*, (19 February 1959): 292.
34 Mills' Methodist Church at Mitcham (designed 1950, completed 1959), with its chapel, hall and classrooms arranged around a courtyard, exemplified his approach to planning for community use. See, Edward D. Mills, *The Modern Church* (London: Architectural Press, 1959) and 'Three Churches: Methodist Church, Mitcham', *AR* (December 1959): 324–6.
35 Wood exhibited two projects at the Church and School Equipment Exhibition (CASEX) at Olympia, St Mary-at-Finchley Parish Halls and Emmanuel Church, Tolworth. KWPP.
36 J.M. Richards, 'Criticism', *AJ* (7 March 1957): 350.
37 On the original plans see *AR* (January 1956): 11–12.
38 On the completed project see *AJ*, (7 March 1957): 350–2; *AJ*, (14 March 1957): 386–7; *AJ*, (16 January 1958): 114–15.
39 Richards, 'Criticism', 350.

40 Richards, 'Criticism', 350–1.
41 'Criticism: The Architect Replies', *AJ* (14 March 1957): 386.
42 'Criticism: The Architect Replies', 387.
43 'Church Hall at Tolworth, Surrey, designed by Kenneth Wood', *AJ* (16 January 1958): 114.
44 'Church Hall at Tolworth, Surrey', 114.
45 *Emmanuel Church Tolworth – This is Your Life*, 10. A history of the church written in 1985. Courtesy of Cara and Colin Rodger.
46 *Emmanuel Church Tolworth*, 38.
47 *Emmanuel Church Tolworth*, 39.
48 'Church Conversion', *A&B* (March 1959): 116.
49 On the American suburban A-frame church see Gretchen Buggeln, 'The Rise and Fall of the Postwar A-Frame Church', a talk given at the Institute for Advanced Study at the University of Minnesota, 7 March 2013, http://ias.umn.edu/2013/05/16/buggeln-gretchen, accessed 10 June 2014.
50 'Timber Gives Church New Look', *Building Materials* (May 1959): 156; 'The Other Cheek', *Interbuild* (March 1959): 24.
51 *AJ*, (19 February 1959): 284.
52 Bridget Cherry and Nikolaus Pevsner, *The Buildings of England*, London 2 South (Harmondsworth: Penguin Books, 1983), 324.
53 Wood used diagonal boarding to the exterior and interior to brace the structure and minimise the effect of any movement. The club incorporated a social area, a sports hall, a workshop for the repair and maintenance of cars, motor cycles and scooters, a dark room and a quiet/committee room. Sliding and folding doors allowed the interior to be configured for different types of activity.
54 Civic Trust Award 1970, KWPP.
55 The Barnfield continues to serve the population of north Kingston as a community centre.
56 Other domestic projects were completed in Wales, Sussex and Essex.
57 *Ideal Home* (June 1961): 49–60.
58 Wood's three-bedroom houses of the late 1950s ranged in size from 1,300 to 1,500 square feet. Span's T range of three-bedroom house types, first employed by Eric Lyons at Hampton Hill in 1952, were generally smaller, ranging from 755 to 1,162 square feet. As a point of reference, F.R.S. Yorke and Penelope Whiting estimated the building cost of a two-storey house of 1,500 square feet at about £3,400 in 1954. Housing authorities were advised to work to between 950 and 1,300 square feet for a three-bedroom house at this time (excluding garages and outbuildings). On Span house sizes see, Ivor Cunningham and Research Design, 'Gazetteer', in Simms, *Eric Lyons and Span*, 189–231. On housing authorities see F.R.S. Yorke and Penelope Whiting, *The New Small House*, 3rd enlarged edition (London: Architectural Press, 1954): 16, 9.
59 Average national house prices rose from £2,230 in 1956 to £4,850 in 1969. See, Christopher Hamnett, *Winners and Losers: Home Ownership in Modern Britain* (Taylor & Francis e-library, 2005), 20.
60 Louise Campbell, 'Patrons of the modern house', *The Journal of the Twentieth Century Society* 2, special issue, 'The Modern House Revisited' (1996): 42–50, on 45, 49.
61 Neil Bingham, 'The architect in society: Eric Lyons, his circle and his values', in Simms, *Eric Lyons and Span*, 11.
62 Kenneth Wood, in discussion with the author, March 2013.
63 Peter Womersley, 'Architects' approach to architecture', *RIBA Journal* (May 1969): 190–6, on 191.
64 Womersley, 'Architects' approach', 194.

65 Dykes (David Drane) and Wildwood (Nigel Money) were also designed for Hawker employees. On John Fozard see, John William Fozard, OBE, *Biographical Memoirs of Fellows of the Royal Society* (1 November 1998): 193–204.

66 'It's twice what it was', *25 New House Extensions and Improvements* (1972): 48–9.

67 'Down go architects commissions', *Design* (February 1970): 18.

68 Kenneth Wood, in discussion with the author, June 2009.

69 Civic Trust, Annual Report, 1973, KWPP.

70 Wood also had regular collaborators. Notable among them was William Willatts, who shared Wood's interest in timber construction and worked with him on the structural side. Willatts trained in engineering and specialised in roof design and the supply of roof products, working for Broderick Insulated Structures of Woking and, later, as an agent for AMFAB – glue laminated structural timber beams, columns, arches, and trusses. Wood may have met him through Eric Lyons or through Kingston College of Art, where he had a part-time position teaching student architects. Paul Willatts, in discussion with the author, September 2014.

71 Jeremy Draper, in discussion with the author, March 2012.

2   The modern house

The modern house in Britain

The context in which Wood was to design private houses in the 1950s began to be shaped in the inter-war period, within a professional arena and through a variety of publications that sought to explain and promote modern architecture to a general audience in relation to the design of their own homes. In Europe, the formal and technical development of modern architecture was closely aligned with the advancement of new ways of living, by Le Corbusier in France, in German-speaking Europe, where it was promoted in publications such as Bruno Taut's *Die neue Wohnung: Die Frau als Schöpferin* (1924) and Sigfried Giedion's *Befreites Wohnen* (1930), and in other countries in which industrial and aesthetic progress was coupled with a new emphasis on 'light, air and openness' in domestic design and the design of public buildings.[1] For Giedion, the interpenetration of interior and exterior space that had started to emerge as a characteristic feature of modern architectural design of the 1920s articulated a new sensibility:

> Today we need a house, that corresponds in its entire structure to our bodily feeling as it is influenced and liberated through sports, gymnastics and a sensuous way of life: light, transparent, movable. Consequentially, this opened house also signifies a reflection of nowadays mental condition: there are no longer separate affairs, all domains interpenetrate.[2]

Although few of Britain's inter-war houses were directly inspired by European avant-garde models, the influence of experimental approaches to the design of the private house can be seen from the late 1920s. The Daily Mail Ideal Home Exhibition, the first domestic exhibition of its type and an arbiter of public taste, introduced British consumers to modern architecture in 1928, with its first 'House of the Future' – a house of radically modern style, with a roof terrace, a roof-top pool and a striking cubist-inspired geometric garden.[3] Several significant books on modern domestic architecture were also published in Britain from the 1920s and show how new approaches to the organisation of domestic space were presented within a framework of continuity, centred on discussions of the plan.

Gordon Allen's *The Smaller House of To-Day* (1926) aimed to present 'modern domestic architecture in its more economical form' and made a case for rationalisation, writing:

> If our new home is to be really satisfactory, it must have a good plan. Elevations, interior effects, in fact the whole scheme, depend on this lay-out ... the plan itself should above all be simple and straight-forward. And when we have discovered the best solution it will appear to be inevitable.[4]

The point is reiterated later in his text: 'Success, whether artistic or economic, can come only where the exterior emerges simply and directly from the plan.'[5] Good light was emphasised in conjunction with health and hygiene: 'The value of sunshine in the right part of the house at the right time of day can hardly be over-estimated, and indeed no rooms are considered healthy unless they periodically receive the disinfection of its rays.'[6] When it came to material expressions of architectural modernity Allen struck a note of caution, advising readers against creating a 'chilling effect' by using too much glass.[7] Rooms, he argued, should be oriented according to patterns of social use and the physical and psychological advantages to users:

> Early morning sunshine in the bedroom has surprisingly beneficial effects on the health and spirits of many people; and a bath-room similarly situated has obvious advantages. Living-rooms used in the morning, then, should have a south-eastern aspect. A drawing-room, or parlour, may be further west. The east side is best for working kitchens, bath-rooms and offices, in order that the earliest risers may benefit from the sun before there is much heat from other sources.[8]

The smaller house was also to be a site of efficiency, stripped of unwanted passages and circulation spaces and with built-in fittings to reduce dust and minimise housework.[9] In such houses compression was the order of the day. Readers were advised of the benefits of a kitchenette, rather than a kitchen – a compact space, designed to accommodate essential activities in condensed form: food preparation and cooking, washing and drying crockery, laundry and ironing and the storage of food and utensils.[10] The living space, Allen suggested, should be arranged as a 'single large living-room, with a "lounge hall," in which to receive visitors whom one may not wish to introduce into the family circle' and a dining recess, linked to the kitchen by a hatch. This was, he explained, a device well-suited to the servantless home where 'the usual drawback to a hatch – that conversation will be overheard in the kitchen – does not apply'.[11] The smaller modern house was, then, to be arranged on a simplified plan, designed to ensure familial privacy and comfort in servantless times.

F.R. Yerbury's *Small Modern English Houses* (1929) adopted a similarly conservative tone in addressing its readers, arguing that a 'tradition of simplicity' distinguished English work from progressive continental architecture and that English architects and clients differed from Europeans in understanding the house as the product of a way of life, rather than a determinant. Yerbury saw no place for the 'very "advanced" type of house' that was being built in Europe and commented: 'Domestic life here still pays respect to human qualities with all their pleasant failings and possibly unscientific prejudices and inconsistencies. Everyday home life is not yet regarded as a science nor is the house regarded as a machine.'[12] In Sweden, which was to have a profound influence on British domestic architecture and interiors of the 1950s, the architect and designer Josef Frank was among those who were beginning to map out an alternative vision for the modern home. Writing in *Form*, in 1934, soon after his arrival in the country: 'the home must not be a mere efficient machine. It must offer comfort, rest and coziness (soothing to the eye, restful to the soul) ... [There are] no puritan principles in good interior decoration.'[13]

F.R.S. Yorke, a founder member of the MARS Group, published a number of books on the design of the modern dwelling between the wars – *The Modern House* (1934); *The Modern Flat* (1937), with Frederick Gibberd; and *The Modern House in England* (1937). *The Modern House*, which ran to several editions, introduced the work of a range of European architects and emphasised new construction techniques and materials as the basis of new domestic forms.[14] In contrast, Raymond McGrath's *Twentieth-Century Houses*, of the same year, included examples of 'good new houses' from twenty countries and considered both the technical and social aspects of their design. The book included contemporary houses from Europe, South Africa, Australia, the United States and Japan, early twentieth-century British examples by C.R. Ashbee, C.F.A. Voysey and Charles Rennie Mackintosh and work from the early 1930s by Serge Chermayeff, Wells Coates, Colin Lucas and Oliver Hill, among others. McGrath made his selection on the basis of 'value of design' rather than price of construction, and most of the houses that were featured cost between £1,000 and £3,000. Intended to provide 'an education in a new way of thought and a wise way of living,' *Twentieth-Century Houses* was written using Basic English, a simplified form devised for international use by C.K. Ogden of the Orthological Institute. In his introduction to the use of Basic English in the volume Ogden drew parallels between McGrath's choice of language and the direct quality of the architecture that his book described.[15]

Like Yerbury, McGrath was concerned with the way in which international architectural developments might be understood in relation to national patterns of domestic life. For McGrath, the 'complete building' not only provided 'cover and comfort' to occupants but also made 'a stage for all their interests and pleasures'.[16] Drawing a distinction between rooms designed for 'the pleasures of living' and those designed for 'the needs of living' McGrath identified certain national characteristics that conflicted with emerging modernist paradigms.[17] In contrast to hard-line modernists' valorisation of the factory kitchen (and while acknowledging that significant work on kitchen design had been completed by Walter Gropius, J.J.P. Oud and Marcel Breuer) he placed the living room at the heart of the house and its planning.[18]

> The most important thing in this room is the seating, from the placing of which heating, lighting and airing take their form – because as the living-room is the centre of the house, so the seating is the centre, or group of centres, of the living-room and will have a necessary amount of view and comfort.[19]

With regard to bedrooms, he felt that limited attention had been given to them in England, where in the middle-class home they remained private retreats:

> The tendency at present is to make the bedroom more of a living-room, sometimes with a writing-table and a small winter garden, and to have it planned wherever possible with a sleeping-terrace, dressing-room and bathroom as a self-dependent series.[20]

McGrath's description points to the continuation of nineteenth-century domestic features as living spaces were architecturally transformed in response to emerging attitudes to healthful living.

Published in 1935, Anthony Bertram's *The House a Machine for Living In* took its cue and title directly from Le Corbusier. J.M. Richards considered it 'a somewhat unwise label to have attached to a popular book whose main purpose is to emphasize the humanitarian case for modern design'.[21] In making his case for a 'new architecture' that was 'logical, clear-sighted, unsentimental, creative', Bertram presented modern architecture as the natural outcome of changing social requirements.[22] Like McGrath, he considered the main elements of the modern plan, including the introduction of more open living environments, and suggested a careful ordering of the interior for different uses, commenting: 'We like to see, even in this room-of-all-work, which part is for what.'[23] As 'the engine of the house-machine' the kitchen was represented as a workspace that 'should never be used for living in if it can be avoided'.[24] Bedrooms required separate dressing-rooms, something that Bertram argued for even when little more than a step-in cupboard could be provided.[25] Space and a sense of spaciousness were the main requirements:

> We want space. Psychologically we want space: to feel at ease, to feel release from the rush and pressure of the outside world, to feel a largeness which our working lives probably do not possess. Physically we

> want space: to move easily about our different occupations, to play games, to move our furniture, to take advantage of the new flexibility and not, like a snail, to be limited to the convolutions of our shell. Therefore the living-room must be as large as possible; its window as large as possible to give our eyes greater space, the furniture in it as small and useful as possible – not three pieces where one will serve, no pieces that do not serve, no pieces that are distorted to a service they were not meant for.[26]

Within these British publications of the 1930s progressive architectural values intersected with domestic ideals and conventions with a longer history, shaped through gendered and class-defined patterns of occupation and in relation to the national ideal of the individual private house and garden.

As thoughts turned to the re-establishment of family life after the Second World War, domestic architecture and design came under intense scrutiny. In 1941 the architect Howard Robertson commented on British developments:

> A few good small experimental houses have been built and lived in, designed by architects who understand the advanced contemporary idiom. They are sometimes marred by eccentricity, such as an overabundance of glass in places where it is not wanted, circular stairs impassable for furniture or a coffin, and surprised looking balconies on which one cannot sit. But they do have an intellectual appeal, and in the alert sanity of their design represent the enlightened left-wing of their category.[27]

In the same year, C.H. Reilly anticipated post-war developments, writing:

> If, then, one may attempt the dangerous role of prophet, it seems likely that the course of domestic architecture after the war, while not returning to the copying of the great traditional styles or even the popular neo-Georgian and Georgian-Swedish compromises of the last decade, will hover between those who build freely, but solidly, in brick and stone and those who build freely, but lightly, in wood or ferro-concrete, rather than between the traditionalists and the modernists. No one to-day wants a reproduction of an old piece of furniture, save for a special purpose, and equally, I fancy, no one tomorrow will want a reproduction of an old building.[28]

Certainly few young architects of Wood's generation wished to build one.

The New Small House (1953), by F.R.S. Yorke and Penelope Whiting, was published just as restrictions on private house-building were coming to an end. The book presented an illustrated collection of houses and housing schemes completed since 1939 and covered work from Britain, America,

Sweden, Denmark and Switzerland. Its authors struck a somewhat pessim-
istic note in their introduction:

> It is surprising to see what some of the eminent people do when faced
> with a small house problem; it is disappointing to find so little adven-
> ture. Not very many of the houses in this book have an appearance
> that expresses a new way of living, or the use of new materials; even
> prefabricated and part prefabricated houses, which have shown some
> technical advances, have little charm, and have not the architectural
> merit that would encourage one to include them.[29]

Even so, they were encouraged by evidence of some improvement in the
design and layout of housing that avoided 'the monotony of the old kind
of two-storey semi-detached suburban sprawl' and by better standards of
planning, with greater consideration being given to the simplification of
housework, the practical location of furniture, the 'intelligent use of new
materials' and improved heating and insulation.[30] Although there had been
a return to a more traditional way of building, the authors noted that prefabricated buildings still accounted for one in five new houses.[31]

Two years after the removal of restrictions on private building *Architectural
Design* reconsidered the status of the modern house as 'a product not so
much of economic conditions as of ideas about architecture and about a
desirable way of life'.[32] A special issue of the journal discussed a range of
international examples, including English projects: two houses at Oxshott
by Powell and Moya, which inspired Wood's clients, the Moneys, to build
nearby (see Chapter 4); a timber and glass weekend house (on concrete
posts) at Bosham, near Chichester harbour, by Architects Co-Partnership;
and Peter Womersley's Farnley Hey, near Huddersfield. The authors noted a
tension in many of the designs:

> At present we are caught in this dilemma: trained in a craft appreciation
> that has almost disappeared, we are hesitant to surrender control to a
> system of prefabrication. Most of the houses illustrated, though some
> are the products of our most sophisticated designers, suffer from this
> schizophrenia. Our architecture is still transitional and we are in no
> position to consolidate any style. We can only select – and explore.[33]

That sense of change was not only evident stylistically, but also in the
space of the interior. In 1959 *Ideal Home* magazine and the RIBA jointly
promoted a small house design competition which attracted over 1,500
entries from qualified architects. In his review of the competition for *The
Builder*, the architect Arthur W. Kenyon was delighted to see no evidence
of semi-detached houses, which had become so closely associated with routine suburban development, commenting: 'If the competition has done no
more than kill the suburban "semi," fifteen hundred architects have earned

a nation's blessing.'[34] Nevertheless, he was somewhat unimpressed with the general standard of the designs submitted, noting a degree of conformity that in his view expressed a lack of creativity:

> Dining rooms are definitely out and the kitchen has become once more a unit of its own, but with probably a hatch breaking its way into something which is vaguely known as a dining space. No one knows quite what to do with this space or where to have it ... The living room is equally uncertain of itself, not knowing whether to plump for comfort or view ... I would say that many of the designs will have little popularity in Tottenham Court Road for free wall space is nil and it must be generally assumed that furniture has to be put in front of windows or cut down to the very minimum required to accommodate human form in repose.[35]

These writings of the 1950s point to many of the uncertainties that were evident in the design of houses of that decade: stylistic uncertainties, tensions between traditional and modern building techniques, and ambivalent responses to open planning. The most competent architects of the period were those, such as Wood, who were able to balance visual and social priorities in line with the opportunities of each site to create liveable environments that satisfied their clients' practical and emotional needs.

The modern house in Surrey

How were these evolving architectural and domestic ideals reflected and shaped within the geographical and social context of Surrey, a county with a rich history of domestic architecture? In his architectural history, *The Surrey Style* (1991), Roderick Gradidge noted the character of domestic development in the county:

> The habit of wealthy Londoners retiring to Surrey, which has now been going on for at least four hundred years, has meant that the county has always been swamped with money. However, because the farm land was poor, the county has rarely appealed to the plutocrat intent on building up a vast estate with a grand house. Instead, it was chosen by the comfortably wealthy – courtiers and merchants – who did not feel the need for broad acres to support their wealth, and it was this gentry class that built the small manor houses that abound in the county.[36]

A number of these old Surrey houses, among them Great Tangley Manor at Wonersh and Rake House at Milford, were altered in the nineteenth century by architects such as Philip Webb and Ralph Nevill.[37] In the early twentieth century, through the work of Edwin Lutyens and Gertrude Jekyll in particular, Surrey became closely associated with the emergence of an idealised domesticity rooted in Arts and Crafts traditions. This homeliest of Home Counties was described by Arthur Haslett in a 1934 article for *The New Estates Magazine*:

> It is the county to which, above all the others, the Englishman can return from overseas, and find the many sided peace of home. It is the county which above all others saves the Londoner from his accursed city, in which if a man should pass three hundred and sixty-five days in a year he would surely die.[38]

Largely rural in nature at the beginning of the twentieth century, Surrey developed rapidly, its population rising from 319,600 in 1901 to 902,200 in 1961.[39] Much of that growth was supported by inter-war house-building, which peaked in the 1930s as construction costs dropped and private ownership became affordable to a wider sector of the British public.[40] Improving road and rail links fostered the outward spread of south-west London's suburbs and the development of new suburban districts around ancient population centres such as Kingston upon Thames.

Proximity to London formed a focus for the marketing of new estates across the Home Counties in the 1930s. Large house-building firms emphasised good transport connections in their publicity materials.[41] In 1935 *The New House* magazine featured a handy table for potential buyers showing '100 districts within easy reach of London' with information on the cost of day return and season tickets.[42] The magazine's 'Survey of London's Dormitories' described several parts of Surrey in which house-building was taking place. Among those were Purley, whose hilly location made for good views and a 'dry and healthy' atmosphere; Esher, with 'all the charm associated with a well-wooded area'; Ewell, 'a quaint and charming village near Epsom'; and Hinchley Wood, whose development had 'been greatly fostered by the proximity of the Kingston by-pass road, and by the opening of a new station on the Southern Railway'.[43] Other Surrey estates that were marketed in the mid 1930s included Carshalton, North Cheam, Sanderstead, Stoneleigh, Surbiton, Wallington and Godstone South.[44]

Surrey house-building firm E. & L. Berg advertised its Berrylands estate with reference to the convenience of Surbiton's new open air swimming pool, The Lagoon, 'one of the finest bathing pools in the country' and its proximity to 'many of the most famous Surrey beauty spots'.[45] Advertising for Bell Property Trust's estate of modern houses, the Parkside Estate at Berrylands, made similar mention of the modern bathing facilities.[46] Other Berg developments, such as the Gladeside Estate adjoining Beckenham Golf Course, promoted aspirational, leisure-oriented lifestyles to middle-class buyers. *Favoured Surrey*, a 1930s brochure for Surrey developer Downs Estates Limited of Banstead emphasised the beauty of the North Downs landscape and the suitability of its soil for 'garden cultivation', recognition of the increasing popularity of recreational gardening as a middle-class domestic pastime.[47] Both firms emphasised the individuality of their houses. Downs Estates promoted theirs as architect designed to individual specification, signalling possibilities for social differentiation and distinguishing their developments from the more uniform estates of other speculative builders.[48] At the time of their construction many of these new estates were in semi-rural locations, which was seen as a major selling point by developers. In 1934 a special Olympia Exhibition edition of *The New Estates Magazine*, edited by Ellis Berg, provided potential buyers with information about the Coombe Estate near Kingston upon Thames, endorsed by Dame Sybil Thorndike as 'Living in the Country made easy and possible!'[49]

A number of large private housing estates were also developed in the first half of the twentieth century. Notable examples are those established by master builder Walter George Tarrant, which included St George's Hill, Weybridge (964 acres, acquired in 1911 and developed from 1912) and Wentworth (developed from 1922). Both had golf courses designed by H.S. Colt and featured houses on plots of at least one acre, often in the Surrey style, 'with tall chimneys, dormer windows, gables, leaded lights, tile-hung or half-timbered or a combination of both'.[50]

Modern and moderne: speculative housing of the 1930s

While many Surrey house-builders drew on local vernacular styles, from the 1930s onwards a number of firms began to test the appeal of aesthetically modern houses. Articles on modern, architect-designed houses were also included in magazines aimed at middle-class home buyers. The October 1935 edition of *The New House*, for example, featured the 'Ultra-Modern' cubist form of Colin Lucas's reinforced concrete house, The Hopfield, at Wrotham in Kent (1931–33).[51] Most significant of the modern, architect-designed estate houses that were built in Surrey between the wars was the Sunspan House. First exhibited at the 1934 Daily Mail Ideal Home Exhibition, the house was designed by Wells Coates and David Pleydell-Bouverie and was licensed to E. & L. Berg Limited for construction. Examples were completed at New Malden (see Figure 2.1), Hinchley Wood and Long Ditton. The Sunspan House was an early attempt to create an economic standard modern house design that could be implemented in an estate layout. In it Raymond McGrath saw the development of 'an international idea ... that of a completely free plan which would make it possible to have the size of the rooms changed whenever necessary by taking space from one and giving it to another'.[52]

Designed to provide maximum light to the interior, the Sunspan House was also planned to offer similar levels of privacy to houses of traditional design, an essential social and symbolic requirement for those moving from more densely populated parts of London, for whom privacy was an indicator of social status as well as an essential constituent of the suburban ideal.[53] *The*

Figure 2.1 A Sunspan House at Woodlands Avenue in New Malden.
Credit: ©Fiona Fisher, 2014.

New Estates Magazine noted the practical advantages of the design, among which were garages with rear doors and 'wash-down' areas to allow cars to be 'cleaned in privacy' – revealing contemporary attitudes to suburban etiquette within these new residential contexts.[54] At New Malden, a mile or so away from Berg's Sunspan houses, Wates and Wates Limited completed a number of modern houses by R.A. Duncan, co-designer of the first 'House of the Future' exhibited at the Daily Mail Ideal Home Exhibition in 1928.

Fashionable styling played a critical role in creating consumer interest and differentiating private estates from the pared down modern designs of local authority housing. A number of developers, among them Curtis Estates of London, a Streatham-based firm, drew on the popular art deco style, advertising 'Sunshine Houses' on its 'Moderne Estate' at Worcester Park.[55] Another of Surrey's modern inter-war housing estates, Upper Farm Estate (1934), by Howard Houses, was built within walking distance of Wood's practice. An advertisement for its 'Ultra Modern Sun-Trap Howard House' emphasised the benefits of the roof terrace – an 'extra outdoor floor' – and the pleasures of open air living: 'There your children can play in safety and unrestricted sunshine. There you can take your meals in the open, entertain your friends, enjoy the peace of moonlight, and sleep al fresco if you wish.'[56] The venture was not a commercial success and despite extensive advertising the developer went bankrupt in 1935.[57]

The architect-designed house

Although the number of modern architects working in inter-war Britain was relatively small, their work was concentrated in London and the south-east of England. Surrey attracted a number of wealthy private clients for modern houses of individual design in this period. Jeremy Gould's 1977 gazetteer of modern houses included over 900 examples, of which fewer than 100 were 'north of Cambridge'.[58] The revised gazetteer, published by the Twentieth Century Society in 1996, identified twenty-nine modern houses at Kingston upon Thames and a further forty-five in Surrey.[59] Among them are two important examples of European-influenced design: Miramonte (1936–37) at Coombe, near Kingston upon Thames, designed by Maxwell Fry for Gerry Green and a few miles away at Esher, The Homewood (1938), designed by Patrick Gwynne for his parents (see Figures 2.2 and 2.3). Both houses were built in mature Surrey landscapes of a type that Wood was to design for after the Second World War, Miramonte in an old park (within walking distance of Wood's Vincent House and Picker House) and The Homewood in the garden that Gwynne's father had designed for the family house at Esher that it replaced.[60]

Miramonte's strongly horizontal design in reinforced concrete calls to mind Mies van der Rohe's Villa Tugendhat at Brno (1930), which may well have been in Gordon Russell's mind when he set out the case for the modern house in a broadcast of 1933:

> Much less time is spent in the house than formerly – our sitting-room is spreading outwards to the loggia, sun-porch, garden, and even to our golf links and open road. Our cars have indeed become our sitting-rooms for quite a long period of each year. Take another example; the large plate-glass windows which wind down into the walls in some modern German rooms emphasising this connection between the living-room and the open air. They frame up a piece of country as if it were a

Figure 2.2 Edwin Maxwell Fry. Miramonte, Kingston upon Thames, London, 1937.

Credit: Architectural Press Archive/RIBA Library Photographs Collection.

picture on the wall, thus bringing it into the room when the window is closed: open it and the room goes out to meet the countryside.[61]

While some evidence of a new spatial consciousness can be seen in the way in which the indoor and outdoor spaces of Miramonte were related through balconies, terraces and the treatment of garden boundaries, the interior maintained a less fluid, largely traditional layout of separate rooms.[62] It is, in that respect, consistent with the spatial conservatism that has been identified as a characteristic of British architectural modernism in the 1930s.[63] The Homewood, which is of marginally later date, is far more open in plan, with a large rectangular first floor living area that was intended 'to combine a background for entertaining with the informality of an all-purpose room'.[64] Drawings of the interior completed for publication in 1938 show the way in which Gwynne defined areas of the living space – for music and cards and for sitting and lounging – through the design of fixed and loose furniture and their placement within the interior and in relation to the exterior landscape.[65] In 1945, when Miramonte was put up for sale, it was promoted by auctioneers Hillier, Parker, May and Rowden as 'A Hollywood House in Surrey'. The auctioneers no doubt hoped that an allusion to American

Figure 2.3 Patrick Gwynne, The Homewood, Esher, Surrey, 1938.
Credit: ©Fiona Fisher, 2013.

glamour would resonate more positively with potential buyers than one to
Miramonte's European origins. This also points to the lead that America
had begun to take in popular culture, architecture and design.[66]

Modern housing of the 1950s

Surrey was also the location in which a highly influential model of modern speculative housing emerged in the 1950s. Span located its modern
housing developments in what Robert Furneaux Jordan described in 1959
as, 'the left-over areas of the more salubrious outer suburbs – Richmond,
Twickenham, Ham, Blackheath and Beckenham – where commuting has its
compensations in sylvan surroundings'.[67] From the late 1940s, Eric Lyons
and Geoffrey Townsend completed a number of projects in Surrey, primarily in Twickenham and Weybridge: Oaklands at Twickenham (1948),
Box Corner and Onslow House, Twickenham (1951), Third Cross Road,
Twickenham (1954), Sandpits Road, Petersham (1954), Campbell Close,
Twickenham (1955), Parkleys, Ham Common (1956), The Cedars, and
Cambridge Court, Teddington (1958), Thurnby Court, Twickenham (1958),

Princes Road, Kew (1959), Victoria Drive, Wimbledon (1960), Fieldend, Teddington (1961), Templemere, Brackley and Castle Green, Weybridge (1965), Weyemede, Byfleet (1966), Holme Chase, Weybridge (1966), Grasmere, Byfleet (1967), Westfield, Ashtead (1969) and Mallard Place, Teddington (1984).[68] The Span housing schemes began to subtly reshape suburban architecture around London's fringe and the distinctive context in which the firm operated led Furneaux Jordan to identify Lyons as 'really a new phenomenon, neither provincial nor metropolitan – a successful suburban architect'.[69]

One of the reasons for Span's success was its targeting of a particular type of moderately progressive middle-class buyer who wanted a house that was modern, unostentatious and practical. Michael Webb, writing in 1969, commented on the success of the Span concept: 'The estates work well because they are tailored to the tastes of a fairly limited social group. The architecture is reticent, relying on attractive grouping, detailing and landscaping to create a harmonious composition.'[70] The spaciousness of the designs was central to their success: 'Internal arrangements are also based on a limited number of types, relying extensively on open plans and staircases, and on folding room dividers. Lighting and other services are exceptionally generous.'[71]

Although modern architecture gained broader acceptance after the Second World War, Surrey's domestic architectural heritage meant that it was not always an easy location in which to build in the modern style and Wood often encountered difficulties in realising his designs. As Ian Nairn noted in his introduction to the Surrey volume of *The Buildings of England* (1962): 'Modern architects still find it more difficult to put up modern buildings in Surrey than almost anywhere else, whilst misinterpretations and malformations of traditional Surrey vernacular multiply without hindrance.'[72] Nevertheless, post-war Surrey remained 'a haven for the discreetly sited private house' and Wood was one of several architects, among them Michael Manser, Powell and Moya, Leslie Gooday, Stefan Buzás, and Peter Womersley, who designed modern houses for Surrey clients in the 1950s and 1960s.[73] Collectively, these houses reflect the breadth of contemporary British architectural practice and ranged from examples of timber construction by Wood and Womersley, to the lightweight Miesian steel and glass houses designed by Wood's fellow Regent Street alumnus Michael Manser and the highly influential and technically experimental house that Richard Rogers designed for his parents in the late 1960s, on the south-west London/Surrey fringe (see Figure 2.4).

Whitewood, Wood's first commission, was planned for a small suburban plot in an area of mainly Victorian character. The single-storey courtyard house was designed to minimise visual impact on nearby residents while ensuring privacy and a favourable outlook for his clients. A similar approach was employed for two later houses, Atrium at Hampton and Tanglewood at Hampton Hill. Their compressed plans and carefully manipulated outlooks can be contrasted with those devised for larger houses in semi-rural settings with attractive open views of the Surrey landscape. Oriel House at

Figure 2.4 Richard and Su Rogers. House and guesthouse for Richard Rogers's parents, 22 Parkside, Wimbledon, London, 1970.
Credit: Tony Ray-Jones/RIBA Library Photographs Collection

Haslemere was built on the site of a former orchard and Nathan House at Oxshott on part of the garden of the Knott Park Estate, which had been sold for development in 1953. Another area of changing character was Coombe, just outside Kingston; Wood built two houses in the area, Picker House (Chapter 10) on part of the landscaped grounds of a mid Victorian mansion that was parcelled for development in 1954 and Vincent House (Chapter 6) nearby. In contrast, Wildwood at Oxshott (Chapter 4) was built on what was to become an exclusive residential enclave of new houses of largely conventional design that grew up when Crown Lands were leased for development in the 1950s. Torrent House (Chapter 9) presented a different architectural challenge, that of a small, disused industrial site, on a riverside plot in an area of significant historical architectural interest.

The social and professional context in which Wood operated shaped the development of his practice. Kingston upon Thames was particularly well-placed to make connections with the design world of London while maintaining its own identity as centre for design. The town had a vibrant art school – one of the few RIBA accredited architecture schools – and was among the first institutions to offer professional interior design training. Wood had links to Kingston College of Art through Noel Moffett and his wife Alina, who were friends. Moffett, an architect and educator, had worked briefly on Serge Chermayeff's house at Bentley Wood in the late 1930s as

Figure 2.5   Kenneth Wood. Office suite for Mary Quant Cosmetics, Franklin National
         Bank Building, Park Avenue, New York, 1973.
Credit: Courtesy of Kenneth Wood.

well as spending time at Burnet, Tait and Lorne. He joined Kingston College
of Art in the early 1950s and taught there until 1970. During that time he
worked in practice with Alina, who was also an architect, specialising in
the field of housing.[74] Wood attended talks at the college, sometimes in the
company of his clients, and the college hosted meetings of the Kingston
upon Thames District chapter of the South Eastern Society of Architects. As
part of this creative community Wood's practice benefited in other ways and
at least one of his employees came to him through the college by personal
recommendation.[75]

Design connections were also made through Kingston's flourishing retail
scene. As well as exhibiting his work in London, Wood organised and participated in local architectural exhibitions at Bentalls department store in
Kingston and at one of the first branches of Terence Conran's Habitat. These,
along with other retailers of modern furniture, such as Trend at Richmond,
were places to which Wood's clients turned in furnishing their new homes.
Kingston's retailers were in turn supported by a vibrant local industrial economy that included design and engineering-led businesses, among them the
Hawker Aircraft Company, and Gala Cosmetics, from which Wood drew a
number of clients.

The close working relationships that Wood and his colleagues forged with their clients for private houses often evolved into longer-term friendships and a number of commissions for commercial projects of different scale resulted. I Iis engagement to design Oxshott Village Centre arose from the house that he designed nearby for the Moneys. Eric Paton, for whom Wood designed Fenwycks at Camberley, commissioned him to design an exhibition stand for The Carpet Manufacturing Company and through Wood's connection with Stanley Picker, Managing Director of Gala Cosmetics, he was invited to design New York offices for Mary Quant Cosmetics (see Figure 2.5). This distinct local network of personal and professional contacts supported Wood's practice.

Notes

1 See, Paul Overy, *Light, Air and Openness: Modern Architecture Between the Wars* (London: Thames & Hudson, 2007).
2 Sigfried Giedion from *Befreites Wohnen* (1929), cited in Hilde Heynan, 'What belongs to architecture? Avant-garde ideas in the modern movement', *Journal of Architecture* 4 (Summer 1999): 129–47, on 135.
3 As Deborah Sugg Ryan has indicated, Tudorbethan remained the main architectural style represented from the exhibition's launch in 1908 until the outbreak of the Second World War. During the inter-war years its response to European architectural modernism was primarily stylistic. See, Deborah S. Ryan, *The Ideal Home Through the 20th Century* (London: Hazar, 1997).
4 Gordon Allen, *The Smaller House of To-Day* (London: B.T. Batsford, 1926), 11, 30.
5 Allen, *The Smaller House*, 64.
6 Allen, *The Smaller House*, 31.
7 Allen, *The Smaller House*, 41.
8 Allen, *The Smaller House*, 32–3.
9 Allen, *The Smaller House*, 50.
10 Allen, *The Smaller House*, 35.
11 Allen, *The Smaller House*, 35, 36.
12 F.R. Yerbury, *Small Modern English Houses* (London: Victor Gollancz, 1929), 5.
13 Josef Frank, 'Rum och inredning' in *Form 30*, 1934, cited in Christopher Long, *Josef Frank: Life and Work* (Chicago: University of Chicago Press, 2002), 198.
14 On Yorke and modernism see, Jeremy Melvin, *F.R.S. Yorke and the Evolution of English Modernism* (Chichester: Wiley-Academy, 2003).
15 Ogden wrote of McGrath, 'In using Basic for his book, he has had in mind something more than the fact, important in itself, that this step would give him an international public. He saw in Basic a language with the same qualities as the buildings he was writing about, and which had, for this reason, a special value for his purpose. C.K. Ogden, 'A Note on Basic English', in Raymond McGrath, *Twentieth-Century Houses* (London: Faber & Faber, 1934), 221.
16 McGrath, *Twentieth-Century Houses*, 33, 34.
17 McGrath, *Twentieth-Century Houses*, 38.
18 McGrath, *Twentieth-Century Houses*, 41–2.
19 McGrath, *Twentieth-Century Houses*, 39.
20 McGrath, *Twentieth-Century Houses*, 38.
21 J.M. Richards, review of *The House: A Machine for Living In* by Anthony Bertram. *The Burlington Magazine for Connoisseurs* (May, 1936): 253.

22 Anthony Bertram, *The House: A Machine for Living In* (London: A&C Black, 1935), 114.
23 Bertram, *The House: A Machine for Living In*, 52.
24 Bertram, *The House: A Machine for Living In*, 52.
25 Bertram, *The House: A Machine for Living In*, 64.
26 Bertram, *The House: A Machine for Living In*, 77.
27 Howard Robertson, 'Domestic architecture and the Second Great War', *Decorative Art: The Studio Yearbook 1940* (London: The Studio, 1940), 11.
28 C.H. Reilly, 'The war and architecture'. In C.G. Holme, ed. *Decorative Art: The Studio Yearbook, 1941* (London: The Studio, 1941), 13.
29 F.R.S. Yorke and Penelope Whiting, *The New Small House*, 3rd enlarged edn. (London: Architectural Press, 1954), 5.
30 Yorke and Whiting, *The New Small House*, 5.
31 Yorke and Whiting, *The New Small House*, 12.
32 'The modern house', *AD* (March 1956): 68.
33 'The modern house', 68.
34 Arthur W. Kenyon, 'Designs for sale: small house competition reviewed', *The Builder* (11 September 1959): 187–9.
35 Kenyon, 'Designs for sale',187–9.
36 Roderick Gradidge, *The Surrey Style* (Kingston: The Surrey Historic Buildings Trust, 1991), 7.
37 Gradidge, *The Surrey Style*, ch. 4.
38 Arthur W. Haslett, 'Beautiful Surrey', *The New Estates Magazine*, 1/3 (1934): 2.
39 Surrey census data 1891–1991. Surrey County Council, http://www.surreycc. gov.uk/environment-housing-and-planning/surrey-data-online/surrey-data-population/population-present-at-census-of-population-1891-1991, accessed 23 September 2013.
40 By the beginning of the Second World War owner-occupied housing was estimated at 32 per cent of the British market. See, Peter Scott, 'Selling owner-occupation to the working-classes in 1930s Britain', Henley Business School, University of Reading, 2004. Accessed at http://www.henley.ac.uk/web/FILES/management/023.pdf on 28 January 2013.
41 Davis Estates, which built new estates around London between the wars, made that relationship palpable by locating a show house at Villiers Street, next to Charing Cross Station. See, Film 8466, The Huntley Archives, http://www.huntleyarchives.com, accessed on 27 November 2013.
42 *The New House* (September 1935): 46–7.
43 *The New House* (September 1935): 28.
44 *The New House* (September 1935): 45.
45 *The New Estates Magazine*, 1/3 (1934): 7, 11.
46 Finn Jensen, *Modernist Semis and Terraces in England* (Ashgate: Farnham, UK and Burlington, VT, 2012), 161–2.
47 Downs Estates Limited, *Favoured Surrey*, (1930s), 1. On gardening see Stephen Constantine, 'Amateur gardening and popular recreation in 19th and 20th centuries', *Journal of Social History* 14/3 (1981): 387–406, on 387.
48 Downs Estates Limited, *Favoured Surrey*, (1930s), 9.
49 *The New Estates Magazine*, 1/3 (1934): 27.
50 Mavis Swenarton, *W.G. Tarrant: Master Builder and Developer*. First published as Monograph 24 by the Walton and Weybridge Local History Society. Viewed online at Elmbridge Museum, http://www.elmbridgemuseum.org.uk/e-museum/?document=300.030.010x1, accessed on 29 January 2013.
51 Baseden Butt, 'An ultra-modern weekend-cottage', *The New House* (October 1935): 16.
52 McGrath, *Twentieth-Century Houses*, 104.

53 As David Kynaston has observed, a range of wartime and post-war sources suggest that privacy was an overriding concern for those imagining domestic life after the hostilities. David Kynaston, *Austerity Britain, 1945 51* (London: Bloomsbury, 2007), 51–2.

54 *The New Estates Magazine*, 1/3 (1934): 4.

55 *The New House* (September 1935): 42.

56 A 1924 advertisement for Howard Houses, cited in Jensen, *Modernist Semis and Terraces in England*, 158.

57 Jensen, *Modernist Semis and Terraces in England*, 159–60.

58 Dean, *The Thirties*, 22.

59 Among them were Raymond McGrath's St Ann's Court at Chertsey; houses by Connell & Ward and by Connell, Ward & Lucas at Grayswood, Redhill, Wentworth and Worcester Park; Jolwynds at Holmbury St Mary and Holthanger at Wentworth by Oliver Hill; The Weald at Betchworth by Ernst Freud; The Yews at Leatherhead by Frederick Etchells; three houses by Maxwell Fry, Ilex at Wimbledon, White Lodge at Bagshot and Miramonte at Coombe, and The Homewood at Esher by Patrick Gwynne. See, Jeremy Gould, 'Gazetteer of modern houses in the United Kingdom and the Republic of Northern Ireland', *The Journal of the Twentieth Century Society* 2, 'The Modern House Revisited' (1996): 111–28.

60 The Homewood was featured in *The Architectural Review* in September 1939. On The Homewood and Gwynne's other pre- and post-war houses see Neil Bingham, 'The houses of Patrick Gwynne', Post-war Houses, Twentieth Century Houses 4, *The Journal of the Twentieth Century Society* (London: The Twentieth Century Society, 2000): 30–44.

61 Gordon Russell from the 3 May 1933 broadcast, 'The living-room and furniture', in *Documents: A Collection of Source Material on the Modern Movement* (Milton Keynes: The Open University Press, 1979), 68.

62 Trevor Keeble, '1900–1940', in Fiona Fisher, Trevor Keeble, Patricia Lara-Betancourt and Brenda Martin, eds. *Performance, Fashion and the Modern Interior: From the Victorians to Today* (Oxford and New York: Berg, 2011), 81.

63 Alan Powers, 'The modern movement', in *Modern: The Modern Movement in Britain*, 13.

64 RIBA Library Drawings Collection, London, RIBA22201.

65 RIBA Library Drawings Collection, London, RIBA22201.

66 Miramonte, Coombe Lane, New Malden, Surrey, first proof, sales particulars for the sale of the house by auction on 31 October 1945, by Hillier, Parker, May and Rowden. Kingston Museum and Heritage Service. For a contemporary discussion see, 'A Surrey House in a Park', *AR* (November 1937): 187–92.

67 R. Furneaux Jordan, 'Span. The spec builder as patron of modern architecture', *AR* (February 1959): 112.

68 See Ivor Cunningham and Research Design, 'Gazetteer', in Simms, *Eric Lyons and Span*, 189–231.

69 Furneaux Jordan, 'Span', 112.

70 Michael Webb, *Architecture in Britain Today* (Feltham, Middlesex: Country Life, 1969), 103.

71 Webb, *Architecture in Britain Today*, 104.

72 Ian Nairn in the Introduction to *The Buildings of England: Surrey* (New Haven, CT and London: Yale University Press, 1971), 74.

73 *The Buildings of England: Surrey*, 78.

74 Dennis Sharp, 'Obituary: Noel Moffett', *The Independent* (20 May 1994), http://www.independent.co.uk/news/people/obituary-noel-moffett-1437249.html, accessed 21 September 2013.

75 Among them Martin Warne, who was suggested to Wood's firm by Noel Moffett.

3 House for an artist
Whitewood, 1958

Whitewood, the first private house that Wood designed, was built on an eighty-by-seventy foot plot at Strawberry Hill near Twickenham in a suburban area of mainly Victorian housing and was designed as a studio house for a fine artist (see Figure 3.1). Artists and designers have often elected to locate themselves within a domestic setting for social and economic as well as practical and intellectual reasons. Their working requirements and personal preferences have, at different times and in different locations, given rise to discrete working environments and to spaces within the home in which household activities were accommodated alongside creative practices of professional and amateur status.[1]

In Britain artists and designers were among the first patrons of modern architecture to consider the ways in which new architectural forms might lend themselves to new possibilities for living and working and to put that into practice in designing or commissioning their own homes.[2] Among those who were drawn to do so in inter-war Surrey were the sculptors Dora Gordine and Frederick McWilliam; Gordine to design her own studio house, Dorich House (1936) at Kingston and McWilliam, to nearby Malden, where he commissioned a house from his friend H.A. Townsend in 1938.[3] The relatively high incidence of architect-designed houses for clients from the fields of art and design that were published in British architectural publications of the 1950s suggests that they remained important patrons of modern architecture after the war. Whitewood is one of a number of modern studio houses built in Surrey in that decade, among which are several examples designed by architects for their own use and others for creative clients, such as the house at Ham that Stefan Buzás created for the artist and designer Margaret Traherne and her husband in 1952.[4]

The studio house is a helpful environment through which to consider some of the broader concerns of modern domestic architecture in the 1950s, particularly the development of multifunctioning, open-plan spaces. In March 1956, *Architectural Design* published a special issue on the modern house in which it discussed the development of open-plan design in the work of inter-war architects and regretted its incorporation in 'stereotyped' and 'emasculated form' within British post-war housing policy.[5] Questioning the

Figure 3.1 Kenneth Wood. Whitewood, Strawberry Hill. The entrance courtyard.
Credit: Courtesy of Kenneth Wood.

ability of open-plan living environments to satisfy contemporary domestic requirements, the journal suggested that: 'Perhaps the answer may be a small parlour, with sofas and a TV, and a large, cheaply constructed space which would do duty for workshop, playroom and parties.'[6] Wood's plan for Whitewood conforms in many respects to that proposition, combining a small, zoned living area with a flexible, creative space in which to work, socialise and play.

Wood's clients for Whitewood were the Jones family, Peter, Vicky and their teenage daughter. Wood first met Peter Jones at the Royal Air Force Air Crew Reception Centre at Regent's Park in London during the war, from which both went on to complete pilot training and to serve in the Air Transport Auxiliary. Their friendship continued after their military service as Wood embarked on his architectural training and Jones began to develop his career as an artist.[7] Wood was in many ways an ideal choice of architect. His own professional and domestic arrangements were spatially enmeshed for much of his working life. At his family's Victorian house three of the first floor rooms were given over to the architectural practice: one room as a studio, one as his private office, and one as a library and administrative space. In a somewhat provisional arrangement a screen on the first floor landing separated the 'office' from the rest of the interior. His experience of fashioning a suitable working environment from a house of conventional cellular plan undoubtedly informed his ability to interpret his clients' needs.

Whitewood was designed as a single-storey house of 1,300 square feet and cost under £3,200 to build. Wood developed a three-sided courtyard plan, with two single-storey brick-built wings linked by a central timber-framed element. The wings created a sheltered, south-facing entrance to the front of the house, which had the best outlook (see Figure 3.2). The timber-framed portion was conceived as a dual-purpose studio/living area and given visual and spatial prominence within the design. As views to the north of the plot were limited the north-facing wall of the studio was designed with clerestory lighting. The south-facing wall, which was largely glazed, included a single solid panel to screen a model's couch from the entrance court. A curtained opening and a glazed storage unit separated one end of the studio/living space from a small workroom and adjoining garage. The workroom contained additional shelves for the storage of art materials and gave direct access to the garden. The rest of the family rooms were located in the opposite wing. From the front door a small hall opened into a rectangular living/dining area with a galley kitchen to one side. A second curtained opening separated the studio/living space from the living/dining area and helped to create a sense of openness and flow within the relatively compact interior. In addition to its role as a professional workspace, the studio was designed to be used for day-to-day activities and entertaining guests. The dividing curtains also lent themselves to another use, as a suitably theatrical backdrop to the puppet theatre performances that the family enjoyed. Three bedrooms (ten by nine foot, twelve foot six inches by nine foot, and eleven foot three inches by seven foot three) and a bathroom and separate cloakroom completed the accommodation in the family wing. Whitewood was occupied by the Jones family from 1958, for almost twenty-five years, and Vicky Jones remembers Wood as having been 'extremely sensitive' to their needs in evolving its design.[8]

Wood was among the many British and international architects who began to develop courtyard and patio plans for smaller post-war houses in the 1950s and 1960s, using them to create a pleasing outlook while affording good levels of privacy to occupants in relatively high-density settings. Gerhard Schwab's *Ein Familienhäuser*, an international study of family houses published in two parts in 1962 and 1966, contained work by a number of British architects – Kenneth Wood, Basil Spence, Peter Aldington,

Figure 3.2 Kenneth Wood. Whitewood, Strawberry Hill. The patio outside the front door, with double doors to the lounge, *c.* 1957.
Credit: RIBA Library Photographs Collection.

Roy Stout and Patrick Litchfield, Ralph Erskine, Peter Womersley and James Gowan.[9] In a review of the book for *Architectural Design*, Michael Manser observed the 'evenness of quality' and 'common denomination of design' that connected the work and commented on 'the high incidence of court-yard houses' among the examples shown.[10] Whitewood is one of a number of small courtyard houses completed by Wood from the late 1950s, among them Atrium (1962) and Tanglewood (1962). Working on a larger scale, he developed a double-courtyard plan to relate the private and staff wings of the Picker House (Chapter 10).[11]

Architect and Building News, which reported on Whitewood soon after its completion, noted that Peter Jones had 'professed no sentimental pre-conceived ideas' about its design, implying a modern sensibility that was reflected in the practical emphasis of his brief for a low-cost, low-mainte-nance house with space in which to paint and to mount and frame work.[12] In line with other houses that Wood completed in the 1950s, Whitewood was designed to require minimal upkeep, to allow its owners to manage much of its decoration and repair without calling on expert help. It reflects Wood's personal commitment to economy of form and materials – the product of his modernist, austerity-period architectural training – and was informed by the ongoing shortage and cost of skilled labour at the time that the house was designed. The low-maintenance ideal that informed Wood's designs of the 1950s can also be situated within the context of a post-war emphasis on family life and the importance of time for leisure and the pursuit of individual hobbies.[13]

To minimise expenditure, a restricted range of materials and finishes were chosen for Whitewood's interior. Walls, in exposed brick or rough single-coat plaster, were planned to require little care. The cost of wiring was kept low through the use of ceiling cords in preference to wall switches and the house had no architraves and limited skirtings. This reduced the number of second fixing items and also minimised dust traps to reduce the burden of cleaning. To avoid the need for regular redecoration, the exposed structural timber elements of the central studio/living space were finished with a clear synthetic sealer, which was also used for the windows and doors and for most of the interior fitments, including the chipboard floor.

As well as featuring in professional journals, such as *Architectural Design*, Whitewood appeared in *House Beautiful* and *Woman*, popular magazines that offered valuable opportunities for publicity to the architect in private practice.[14] Reading across these sources reveals the ways in which Whitewood's design was interpreted for different audiences, as well as the weight of authority that was afforded to client, architect and journalist within these different editorial environments. The architectural journals in which Whitewood appeared represented it as a house for an artist and his family, focusing on the formal, functional and technical dimensions of Wood's design. *Woman* magazine, in contrast, published the house under the title 'At Home with Us'. A Five Point Plan for Peter and Vickey' [sic] (see Figure 3.3). Considering the house primarily from the woman's point of view, it concentrated on the ways in which Wood's design fulfilled Vicky's domestic requirements and personal aspirations.

A close relationship can be found between certain passages of text that appeared in professional and popular sources, suggesting that they derived at least some of their content from material generated by Wood's office. *Architect and Building News*, for example, described Whitewood as 'a house that would be unselfconscious and which could develop its character over the years, providing a background, not a straightjacket, for living and painting'.[15] A similar view was expressed in *House Beautiful*, which commented: 'But most important it was to be a house that would be un-selfconscious and could develop its character over the years – a background, not a strait jacket, for living and painting.'[16] Rather than expressing an editorial opinion, these revoiced quotations articulate Wood's values and the ideal of unobtrusive domestic efficiency that informed his approach to the design of modern living environments. Also evident in the two descriptions is an accent on stability of tenure that appears in many respects antithetical to those ideals of mobility and

Sunny courtyard
Peter and Vickey built their house round the trees on their 80 ft. by 70 ft. plot. The two brick wings linked by the combined studio and living-room shield a courtyard to make warm sitting-out spot

AT HOME WITH US

EDITH BLAIR, Home Editor, takes you into a modern house that's designed to be light, warm, easy to run, low in running costs, and a simple family background

FIVE
POINT
PLAN

for Peter and Vickey

PETER and Vickey Jones drew up a five point house plan for their architect to work on when they had bought (with a small legacy) a £700 plot of land. The plot is at Strawberry Hill in Middlesex, about half an hour's drive from the centre of London.
"I have a horror of being boxed in by walls," Vickey said. " I asked for lots of window wall, and we all dislike cutting down trees so we pleaded that the lovely fruit trees on our plot should be preserved."
Peter is an artist and picture restorer so his demand was studio space as well as plenty of light. Vickey designs fabric pictures which are bought by the large stores, likes working beside her husband, which meant the studio had to be big enough for two.
Housework is not a thing Vickey is mad about, so quick, easy home running was a big priority.
Not many of us are lucky enough to own a home that fits our individual needs but we can often groom one to fit us more closely, and that is why this story of a house is full of interest.
Peter and Vickey got the home they wanted for themselves and their thirteen-year-old daughter Barrie. It's a modern house made up of two

please turn to next page

Studio for two

Colour in the studio is muted to show off their pictures. Chest of drawers from their old home finds a place in the new and fits in happily with its black drawers and contrasting white framework

Restful corner

Special seat was built into corner of studio living-room. Vickey finds it a good spot for comfort doing the family mending

Figure 3.3 Kenneth Wood. Whitewood, Strawberry Hill, from Edith Blair, 'At Home with Us', *Woman*, 4 October 1958.
Credit: Courtesy of IPC Media.

freedom that had inspired leading modernist architects, such as Wells Coates, in their approach to the design of modern housing for the professional classes in the inter-war years.

Other similarities between published materials can be found, foregrounding those elements of the design upon which Wood placed most importance and to which he wished to draw attention. The nocturnal appearance and experience of the building is one example. In *Woman* magazine Vicky Jones is quoted as saying: 'At night when the curtains are drawn we don't get that closed-in feeling; we can see the night sky through our clerestory windows.'[17] Another contemporary article commented: 'Upper shallow windows give reflected light off the ceiling and prevent the usual "dead effect" at night when curtains are drawn.'[18] Both references reveal Wood's concern with the animation of the interior and express a more widespread contemporary architectural preoccupation with the mediation of the domestic boundary and the incorporation of landscape into the interior. While this has been particularly associated with the ubiquitous picture window, clerestory lighting represents an equally important means through which architects sought to preserve a connection with nature in locations with a poor outlook. Examples can be found in Wood's work and that of his contemporaries. In urban and suburban locations, where clerestory windows were used to frame partial views of trees, sky or suburban skylines, without exposing occupants to scrutiny or unattractive views. In smaller post-war houses, where the incorporation of adequate storage was often a challenge, clerestory windows offered a further benefit, providing usable wall space to accommodate loose or fitted furniture.

Whitewood's combination of domestic and studio space gave rise to a variety of contemporary readings. *Architectural Design* and *The Architectural Review* represented the central architectural element of the house as a studio.[19] *Woman* magazine, in contrast, identified it as a 'studio-living room' and emphasised the social elements of Wood's design (see Figure 3.4). These different interpretations were determined as much by the visual approach of the magazines as by their texts. *Woman* included a set of photographs, some in colour, showing details of the interior with the family in occupation. *Architectural Design* and *The Architectural Review* included the same three images – two of the exterior of the house and one of Peter at work in 'his' studio, with several of his pictures on display. These tended to suppress the domestic function of the space, placing it within an established category of discrete studio environments designed for individual creative practice and self-presentation. *House Beautiful* gave greater attention to the design of the interior and included two photographs of the studio/living space, one of the kitchen, and one of the exterior of the house, none of which included the owners.[20]

In its dual emphasis on design and social use and its approach to the photographic representation of the interior and its occupants, *Woman* magazine offered a more complex interpretation of Wood's design than the accounts that appeared in professional architectural journals and the more upmarket, female-oriented context of *House Beautiful*. Following the appointment of its editor, Mary Grieve, to the Council of Industrial Design in 1952, *Woman* magazine played a major role in communicating modernist ideals to non-professional readers through its depiction of real homes and interiors and their framing within a modernist paradigm.[21] In the case of Whitewood, this can be seen most clearly in the magazine's discussion of materials in relation to domestic labour and in its description of the kitchen. 'Housework',

Figure 3.4 Kenneth Wood. Whitewood, Strawberry Hill. Detail of the studio/living
area from *Woman*, 4 October 1958.
Credit: Courtesy of IPC Media.

it claimed, 'is not a thing Vickey [sic] is mad about, so quick, easy home running was a big priority'.[22]

> Vickey [sic] finds the all-over-the-house chipboard flooring wonder-
> fully easy to keep clean ... On this there are luxurious white looped
> pile rugs made by Granny Jones. No paintwork cleaning, no surround
> polishing in this home – and Vickey's friends are green with envy at her
> chore free life.[23]

Author Edith Blair's discussion of the decoration of the kitchen – its Douglas
fir ply fittings and cheerful colour scheme of white with strong red and yellow accents – was given equal weight to discussion of the functional elements of Wood's design. The kitchen was described as 'step-saving' by virtue
of its narrow form.[24] In the post-war period the term 'step-saving' was not
only used to describe rationally planned kitchens, but also began to be used

as a more general justification for kitchens with insufficient space to accommodate the informal styles of dining that emerged as a middle-class preference at this time.

Domestic technology, a focus for post-war writings on architecture and the home, was also discussed in relation to Whitewood's plan. The house was designed to high thermal insulation standards and was heated by a coke-fired warm-air heating system, with grilles in every room. The return duct was concealed within a central core of cupboards in the family wing, optimising space and reducing the need for loose furniture items. Edith Blair commented that the only chest of drawers that she observed was one in the studio/living space, in which Vicky stored the swatches that she used to make her fabric pictures, alerting readers to Vicky's creative use of the space.[25] Vicky sold her pictures through London department stores, but of the contemporary sources in which the house appeared only *Woman* magazine acknowledged her dual role as a housewife and artist.

Magazines such as *Woman* played a central role in promoting an image of the modern family in which gendered roles were clearly defined and its photographs of Whitewood reveal other aspects of Wood's design for this working home that are overlooked, or disregarded elsewhere. Three of the magazine's photographs show Peter and Vicky together. In one they are relaxing in the entrance court. In another Vicky is seated at a trestle table in the studio/living room, working on one of her pictures, while Peter stands to one side at his easel (see Figure 3.5). In the third, Vicky is sitting in the *huff hole* – a built-in seat that was mischievously introduced to the design to commemorate her occasional loss of good humour at points during the construction. Here, she is shown doing her 'family mending' as Peter leans toward her through an open window; an alternative framing of the studio/living room as a site of domestic labour. The magazine's idealised images of a companionate marriage, in which the couple lived and worked side-by-side, suggest ways in which the design of the interior supported their roles within the home and reveals the gendered hierarchies of professional and amateur artistic production inscribed within it. In contrast to the storage fitting that housed Peter's materials, which was designed as an integral architectural element of the interior, items used by Vicky – a folding table and an old chest of drawers – were both movable and of domestic origin.

A clearer sense of the multipurpose nature of the studio/living space can be gleaned from the utilitarian and decorative objects that it contained – house plants, a table lamp and a soft toy or puppet – alongside everyday items put to new use, such as the old chest of drawers and a mug containing Peter's paint brushes. These repurposed items expressed the separate identity of the studio/living space, differentiating it from the family wing while retaining something of its domestic ambience. The adjoining living area was furnished in a more single-minded domestic fashion, with a fitted wall unit, a two-seater sofa, easy chairs, and a side table in contemporary style.

Figure 3.5 Kenneth Wood. Whitewood, Strawberry Hill, from Edith Blair, 'At Home with Us', *Woman*, 4 October 1958.
Credit: Courtesy of IPC Media.

A low open shelving unit divided the seating area from a dining zone to its rear and was used by the family to display decorative glass and ceramics (see Figure 3.6).

Early photographs of the interior indicate a number of artistic additions to the simple architectural background that Wood created. In the kitchen a glazed larder was decorated with Peter's hand-painted pictures of food and drink. Cut out photographic images also brought colour to the bottom of the stable door that led to the garden. The garage wall, which defined one side of the entrance courtyard, was decorated with a ceramic mural

Figure 3.6 Kenneth Wood. Whitewood, Strawberry Hill. Detail of the living/dining
area from *Woman*, 4 October 1958.
Credit: Courtesy of IPC Media.

that Peter Jones and Wood completed together, for which Jones had col-
lected pieces of broken crockery. Both were interested in the place of art
within architectural design and pursued that interest in later projects. Wood
painted throughout his career and went on to incorporate work by other
artists into his designs as well as creating work of his own, including a set
of murals for the Physiology Department at Guy's Hospital in London and
a weather vane for St Paul's School at Kingston.[26] Jones followed his interest as a member of the organising committee that developed the 'Artists and
Architecture' exhibition that was held at the Building Centre Trust gallery
in Store Street in 1967, which was put together with the assistance of the
Arts Council of Great Britain and staff at *The Architectural Review*.[27] The

exhibition was intended as 'a comprehensive assessment of the development and understanding between artist and architect; especially the relationship towards new materials and their application to structures'.[28] Jones submitted 'spaceplace', the first of a number of 'abstract walk-through spaces' completed with Maurice Agis, his professional partner.[29] Also included in the exhibition was a relief construction for the entrance porch of the Forestry Commission complex at Santon Downham in Suffolk, a large commercial project by Wood's firm, to which he had invited Jones to contribute.[30]

In his design for Whitewood, Wood attempted to create an environment in which the studio was neither an adjunct to the domestic, nor the domestic an extension of the studio. The dual function of the house was expressed structurally and materially, in the relationship between the lightweight timber element – a creative space for work and play at the heart of the house – and the adjoining brick-built wings, which provided a more conventional and solid grounding for family life. Wood's use of brick and timber, which emerged as a characteristic feature of his domestic work, reflects the wider acceptance of the use of mixed materials among British architects of the 1950s.[31] In a number of later projects he combined timber and brick to similar effect, using heavy and light elements to relate fixed and flexible spaces within the plan.

Notes

1 Since 2002 the Workhome Project has sought to provide a systematic account of the history of combined living and working environments. See, http://www.theworkhome.com.

2 Paris was an important context in which the modern studio house evolved. See, Reyner Banham 'Ateliers d'artistes. Paris studio houses and the modern movement', *AR* (August 1956): 75–84.

3 On Gordine see, Brenda Martin, 'Four studio-houses: a negotiation of modernism', in Jonathan Black and Brenda Martin, *Dora Gordine: Sculptor, Artist, Designer* (London: Dorich House Museum in association with Philip Wilson Publishers, 2007), 163–205.

4 On architects' houses see, for example, Alan and Sylvia Blanc's Uphill House (1957–60) at Kingston in Miranda H. Newton, *Architects' London Houses* (Oxford: Butterworth Architecture, 1992), 26–31.

5 'The modern house', *AD* (March 1956): 68.

6 'The modern house', *AD* (March 1956): 68.

7 Jones was associated with a group of avant-garde artists that developed around Jack Bilbo and had his first solo exhibition in 1946 at Bilbo's Modern Art Gallery in London. In the early 1960s he worked for several years for Eric Estorick. Alexa Jones, 'Peter Jones', Obituary, *The Guardian*, 22 August 2008. Available at http://www.theguardian.com/artanddesign/2008/aug/22/1, accessed on 8 August 2013.

8 Vicky Jones, in discussion with the author, July 2009.

9 Gerhard Schwab, *Einfamilienhäuser 1–50* and *Einfamilienhäuser 51–100* (Stuttgart: Deutsche Verlags-Anstalt, 1962 and 1966).

10 Michael Manser, *Ein Familienhäuser*, Book Review. *AD* (December 1966), loose clipping, KWPP.

11 The Turn, Middle Turn and Turn End (1963–68), the group of three modern village houses that Peter Aldington designed at Haddenham in Buckinghamshire, are a good rural example. Aldington has stated that he drew inspiration from the innovative approaches of pioneering modernist architects as his work evolved from Askett Green village house 'a box which contains the structure' to the free-flowing spaces of Haddenham. See, Peter Aldington, 'Architecture and the landscape obligation', Post-war Houses, Twentieth Century Houses 4, *The Journal of the Twentieth Century Society* (London: The Twentieth Century Society, 2000): 20–8, on 21. See also, Alan Powers, *Aldington, Craig and Collinge* (London: RIBA, 2009) and Elain Harwood, *England: A Guide to Post-War Listed Buildings*, 2nd rev. edn. (London: Batsford, 2003), 410.

12 'Studio House, Twickenham', *A&BN* (20 May 1959): 657–8.

13 A variety of popular publications such as *Practical Householder* (1955) and *Do It Yourself* (1957) launched in the mid 1950s to cater to householders' needs. As Andrew Jackson's analysis of these titles has shown, their editorial emphasis gradually shifted as early practical and economic concerns gave way, by the early 1960s, to an interest in do-it-yourself and home decorating as means of self-expression and domestic improvement. See, Andrew Jackson, 'Labour as leisure: the Mirror dinghy and DIY sailors', *Journal of Design History* 19/1 (2006): 57–67, on 60.

14 *AD*, 28/9 (September 1958): 360–77; *AR* (October 1958): 263–4; *Woman* (4 October 1958): 31–2; *A&B*, (20 May 1959): 657; *House Beautiful* (April 1960): 56–7; *Financial Times* (13 November 1961): 40.

15 'Studio House, Twickenham. Architect: Kenneth Wood', *A&B*, 215/20 (20 May 1959): 657.

16 'House at Strawberry Hill', *House Beautiful* (April 1960): 57.

17 *Woman* (4 October 1958): 32.

18 'A modern studio home', loose cutting. KWPP.

19 *AD*, 28/9 (September 1958): 360–77; *AR* (October 1958): 263–4.

20 *House Beautiful* (April 1960): 56–7.

21 Trevor Keeble 'Domesticating modernity: *Woman* magazine and the modern home', in Jeremy Aynsley and Kate Forde, eds. *Design and the Modern Magazine* (Manchester: Manchester University Press, 2007), 95–113.

22 *Woman* (4 October 1958): 31.

23 *Woman* (4 October 1958): 32.

24 The concept of the step-saving kitchen originates in the work of American home economists Lillian Gilbreth, Christine Frederick and others, who, in the early twentieth century introduced the scientific efficiency of the factory to the home. See, Sarah Leavitt, *From Catharine Beecher to Martha Stewart: A Cultural History of Domestic Advice*, Chapel Hill: University of North Carolina Press, 2002. Quotation, *Woman* (4 October 1958): 32.

25 *Woman* (4 October 1958): 32.

26 Wood's wall panels for Guy's Hospital were exhibited at the Royal Academy Summer Exhibition and were reviewed in *La revue moderne*, which appreciated his decorative sense and described the panels as being worthy of the best 'op art' painters. 'En Grand Bretagne. Salon de la Royal Academy. Kenneth Wood', *La revue moderne* (1 September 1969), 28.

27 Artists and Architecture 67. Exhibition catalogue. The Building Centre Trust, 1967. KWPP.

28 British work included murals, mosaics, floor and wall panel designs, windows, reliefs and wall sculptures, entrance gates and grilles, screens and trellises. Participants included Mary Adshead, Edward Bawden, Geoffrey Clarke

and Theo Crosby. See, Artists and Architecture 67. Exhibition catalogue. The Building Centre Trust, 1967 and *Design* (November 1967): 22.

29 Spaceplace was exhibited at the Museum of Modern Art in Oxford in 1966 and at the Stedlijk Museum in Amsterdam, where it was awarded the Sikkens Prize. See: 'Art obituaries. Maurice Agis', *The Telegraph* (16 October 2009).

30 Artists and Architecture 67. Exhibition catalogue.

31 On materials see Powers, *The Twentieth Century House in Britain*, 109.

4 Flexible house
Wildwood, 1958

Wildwood at Oxshott was designed for a professional couple who aspired to create a flexible living environment to meet their long-term needs (see Figure 4.1). Although flexibility was well-established as a general principle of modern domestic architecture by the 1950s when it came to articulating their requirements Wood's clients held different views of what that might mean to them in practical terms. The free-flowing plan that Wood developed for Wildwood reflects a different vision of flexible living to that which underpinned Whitewood's design. Central to its expression was the relationship between functionally determined fittings, movable elements, and loose furnishings that allowed the adjustment of the interior for different social purposes. This combination of architecturally determined and user-determined features is characteristic of Wood's interiors of the 1950s and suggests his assimilation of the two principal models of flexible planning that had emerged in the inter-war period.[1] Another sense in which the design can be understood as flexible is the capacity that it had to accommodate minor additions and alterations by its owners in response to their occupation of the space.

Wood's clients, Nigel and Elisa Money, an aeronautical engineer and a mathematician, had worked in Canada in the early 1950s. Young, university educated and interested in contemporary design, they were arguably more adventurous and cosmopolitan in outlook than many of their peers. Both were keen to explore the possibility of creating, rather than buying their own home, perhaps in response to their experience of Canada's strong self-build culture.[2] Soon after their return to England, in 1955, they set about finding a plot of land on which to build.[3] Weekend drives around Surrey eventually led them to Oxshott, where they came across Milk Wood and Headlong Hill (1955), a pair of newly completed houses by Philip Powell and Hidalgo Moya for civil servants Howell Leadbetter and Desmond Keeling.[4] As unusual examples of modern design they were of immediate interest to the Moneys, who called in at Headlong Hill, curious to discover how the owner had acquired the land and obtained permission to build.[5] Furnished with the news that the Crown was selling leasehold plots for development, the Moneys went on to find their own site on the Four Acre

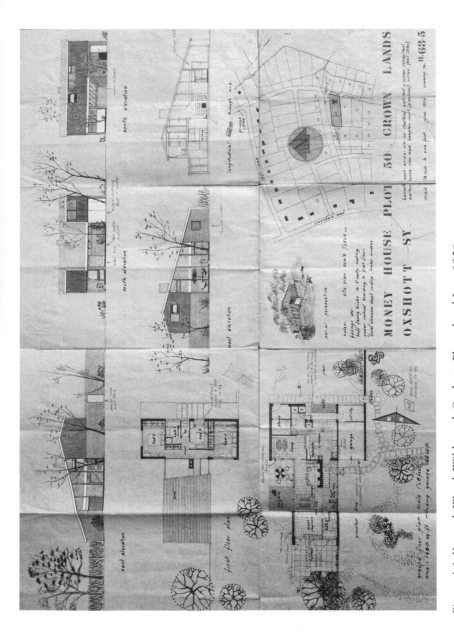

Figure 4.1 Kenneth Wood. Wildwood, Oxshott. Plans dated June 1956.
Credit: Courtesy of Elisa Money.

Plantation Development at Oxshott, signing their lease in February 1957.[6] The terms of the lease were such that houses on the new estate were to be erected at a minimum cost of £4,000, a sum which was to include the construction of a garage and the layout and planting of a garden. Architectural plans were to be submitted to the Local Planning Authority and to the Lords Commissioners for approval.

The Moneys had a clear idea of what they wanted from their new house: a modern, spatially efficient design that would make the most of the attractive woodland setting that they had found. Both the Moneys were keen ornithologists, therefore views from the interior were an important consideration. They had no architect in mind for their project, but were regular readers of *The Architectural Review* and favoured a moderate, Scandinavian-influenced modernism with which Wood was largely in sympathy. His commission came about through a chance conversation with Elisa Money at a local council office, where she was making enquiries about building, and led to an invitation to present the couple with some initial ideas.[7]

Wood first envisaged Wildwood as a single-storey house, but it proved impossible to obtain planning permission for a single-storey building on the site.[8] Alternative two-storey designs were devised, one of which had immediate appeal and was selected just as the Moneys were finalising the acquisition of their lease.[9] Elisa Money was not greatly involved in the design process. She recalls that it was mainly her husband and Wood who conversed on the project; both were 'detail men', who shared an aesthetic vision for the house and an understanding of the language of modern design that allowed easy communication between them.[10] Her personal recollection of the position of women in the 1950s, even university educated women such as herself, was that their views on such matters did not count very much at that time.[11] During the construction of the house the couple lived 'very frugally' and used her salary to cover day-to-day living expenses while his went into the project.[12] They moved into Wildwood in June 1958.

Wood designed Wildwood to be set well back on the plot, surrounded by existing trees and approached by way of a curving gravel driveway. Arranged on a T-shaped plan, the house was slightly larger than Whitewood and provided 1,500 square feet of accommodation including the garage (see Figure 4.2).[13] It was designed with three bedrooms and the possibility of creating a further bedroom for occasional use on the ground floor. Typologically, it was conceived as a galleried house and was configured around a double-height, semi-open-plan living space. Although a number of British examples of galleried houses appeared in publications of the interwar period, it was not until the 1950s that British architects began to exploit the spatial potential of galleried forms more seriously.[14]

Structurally, Wildwood was built using a combination of load-bearing brickwork and glazed timber post-and-beam framing, with insulating blocks to the upper floor. These were clad in Western red cedar weatherboarding,

Figure 4.2 Kenneth Wood. Wildwood, Oxshott. The front entrance and garage, 1959.

Credit: RIBA Library Photographs Collection.

an appropriate choice of materials for a house in a woodland setting.[15] The highly glazed front elevation helped to relate the house to its wooded plot, allowing the reflections of trees to play on the facade where they merged with partial views through the interior to the garden and landscape to the rear.

Wood's preferred contractor declined to tender for the job as he was unsure how to manage the post-and-beam construction without specialist equipment. This difficulty in finding a local builder, for what was then seen as a non-standard construction method, is indicative of the strength of Britain's brick building tradition at the time. In the end, the use of ladders sufficed and the construction was completed without undue drama.

Wildwood's ground floor rooms were arranged as a set of inter-linking spaces, oriented around a central fireplace/staircase wall that separated an open entrance area from a living space to the rear (see Figure 4.3).[16] The fireplace/staircase wall was the dominant feature within the interior and incorporated open shelves to accommodate books or decorative objects. High on the wall, a small slate shelf, which was designed to house a goddess of the hearth, drew the eye to gallery above. The hearth, also in slate, extended

Figure 4.3 Kenneth Wood. Wildwood, Oxshott. View of living room and staircase, 1959.
Credit: RIBA Library Photographs Collection.

the length of the fireplace wall. The galleried living area was designed for flexible use. A folding timber wall allowed its arrangement as one large living area, a living area and adjoining playroom or, in conjunction with the ground floor cloakroom/shower room, as an occasional guest bedroom suite (see Figure 4.4).

In plan, the design suggests Wood's assimilation of the model of rotational planning developed by Frank Lloyd Wright in the 1890s.[17] Wood initially came to Wright's work through his studies. Sigfried Giedion's *Space, Time and Architecture: the Growth of a New Tradition* was incorporated into the architectural syllabus at Regent Street soon after its publication in 1941 and remained an influential source on Wright's work during Wood's architectural training.[18] Giedion saw in Wright's work the development of an American domestic architecture steeped in a tradition of simplicity, flexibility and informality, expressed in his use of materials – 'The plane surface – the flat wall of wood, brick, or stone – has always been a basic element in American architecture' – and in his development of floor plans that could be enlarged according to social need and economic circumstances.[19] Later, on a study visit to North America just before Wildwood's completion, Wood was

Figure 4.4 Kenneth Wood. Tanglewood, Hampton, 1962. View from the living room
to the dining zone.

Credit: Courtesy of Kenneth Wood.

able to experience Wright's work for himself and to observe its influence on
the progress of modern domestic architecture on Canada's West Coast (see
Chapter 5). His assimilation of Wright's architecture is evident in a number
of other projects; in the oriental influence that can be observed in his design
for the Picker House (Chapter 10) and his visual approach to the planning of small houses, such as Tanglewood, to create 'borrowed' and multiple
interior views (see Figure 4.4).[20]

Wildwood was designed with no lobby or hall as such, but with an
entrance space that incorporated the void beneath the staircase and led to

the living area at one side and the kitchen at the other. Open-tread staircases became popular as architects sought to create a sense of spaciousness in smaller post-war homes. As H. Dalton Clifford's *New Houses for Moderate Means* of 1957 indicates, unlike the picture window, a recognisable visual marker of domestic modernity by that time, the open-tread staircase was relatively novel interior feature and required introduction and explanation. The author emphasised its visual and practical advantages over the compromise that it involved to storage, writing:

> Many of the houses illustrated have open-tread stairs, that is to say that, like step-ladders, there are no 'risers' filling the gaps between the treads. They take up the same space as ordinary stairs but look far less substantial because one can see beyond them. Although there can be no cupboard under such stairs the gain in apparent roominess is worth the sacrifice. Furthermore, they require no stair carpet and are much easier to clean.[21]

Wood incorporated extensive built-in storage throughout the house. Built-in furniture evolved as an essential feature of the modern interior of the inter-war period, the relationship between what was fixed and what was loose informing its appearance and spatiality. By the early 1930s, in response to the dissemination of Le Corbusier's work, in particular, architects had begun to reconsider the relationship between architecture and furniture and to think in terms of 'equipping' rather than 'furnishing' domestic spaces.[22] In a 1933 broadcast 'Modern Dwellings for Modern Needs' Wells Coates claimed:

> Very soon it will be realised that it is as fantastic to move from house to house accompanied by an enormous van, filled with wardrobes, chairs, tables, beds, chests and whatnots, as it would seem today to remove the bath and heating system complete including all the pipes.[23]

In the 1950s the use of built-in elements evolved within British houses of traditional and modern form in conjunction with light, easily movable furniture items that were suited to small living spaces and in certain contexts posited an idealised indoor/outdoor lifestyle derived from North American models.[24] Those architects such as Wood, who were most attuned to the needs of their clients, varied their approach to interior design according to the particular circumstances of each project.[25] Nevertheless, certain general observations can be made about the type of fittings that Wood generally incorporated and the ways in which they regulated social use of the interior. Storage was principally built-in and included kitchen fitments, shelving units, wardrobes and dressing tables. Elements that related to particular working requirements were often fixed, signalling and determining a particular use of those spaces. Built-in seating was introduced only occasionally. Banks of floor-to-ceiling

Figure 4.5 Kenneth Wood. Wildwood, Oxshott. View to the living room from the
gallery, 1959.
Photographer: Douglas Whittaker. Credit: RIBA Library Photographs Collection.

wall cupboards were sometimes used in place of walls, allowing the reconfiguration of bedrooms and the concealment of domestic services. Although
Wood was never reticent about giving his opinion on furnishing and decoration, he usually left the choice of such items to his clients.[26]

Wildwood featured in several publications, among them *Country Life*,
The Financial Times, *Architecture and Building* and *House and Garden*.[27]
An article in *Woman* magazine came about through Elisa Money, who had
been at school with a member of the magazine's staff, Patience Buckley.[28]
In it the magazine's editor, Edith Blair, described Wildwood as 'a house that
really belongs to NOW' and in a brief description detailed its attractions,
among which were its light-filled interior, its economic spatial planning,
its low-maintenance features and built-in furniture.[29] The fitted cupboards
in the bedrooms and dressing table in the master bedroom were especially
appreciated as 'good ideas to copy to make small-room living easier'.[30]

Wood designed a number of fitted elements for the main ground floor living areas, including a small study space, with an integrated hall cupboard
that separated it from the entrance area (see Figure 4.5). Larger fitments
in the kitchen and dining room played an important role in shaping the

Figure 4.6 Kenneth Wood. Wildwood, Oxshott. The dining room fitment, 1959.
Photographer: Douglas Whittaker. Credit: RIBA Library Photographs Collection.

visual and social experience of the interior. At one end of the dining room a
floor-to-ceiling display and storage unit filled the end wall. Finished in syca-
more, teak, Formica and glass, it contained open shelves for books, glazed
shelves for the protected display of delicate articles, such as glass and china,
and plenty of cupboards to allow the concealed storage of other items (see
Figure 4.6).[31] Photographs taken shortly after Wildwood was completed
show the dining room unit filled with books and objects of modern design
including simple, unornamented glassware, the creamer from a Raymond
Loewy-designed Rosenthal tea set, rush table mats and (what appears to be)
a wooden monkey designed by Kay Bojesen. These utilitarian and decorative objects, a number of which were bought by the Moneys at Heals department store in London, sat alongside items of older origin, tankards and an
ornamental sweet dish that evoked more tradition domestic settings.

Wood's built-in furniture was complemented by modern, lightweight furniture chosen by his clients. For the living area the Moneys selected furniture by Ernest Race and Hille that could be oriented to the garden or turned
toward the fire in cooler weather. The Race 'Heron' chairs that appear in
published photographs of the interior were 'very elegant' but not so good
to sit on according to Elisa Money; they were her husband's choice and one

made, in her opinion, with an eye to aesthetics rather than comfort.[32] They were disposed of in time, but a dining table by Terence Conran outlasted them and has remained in use for over fifty years.[33]

Another notable constituent of Wildwood's flexible interior was the use of curtains, an element of interior design claimed by modern architects as part of the essential 'equipment' of the modern house.[34] Representations of contemporary architect-designed houses suggest a variety of spatial and decorative, client and architect-led approaches at play in their use. In her design for a house at Hyver Hill, Hendon, for Mr and Mrs Stanley Broadbent, Jane Drew used curtains as a spatial divider within the interior. The house benefited from a good view and Drew incorporated floor-to-ceiling windows the length of its thirty-three foot long living room wall. Curtain tracks were designed to allow the curtains to be drawn clear of the windows and either 'draped along the back wall of the dining-recess' or used 'in an intermediate position to screen the recess from the rest of the living room'.[35] Drew's approach to the use of curtains to define interior space and frame an exterior view can be contrasted with that of textile designer, Tibor Reich, who made extravagant use of curtains in the interior of the house that Denis Hinton designed for him, for which Hinton introduced double curtain tracks to the living room to allow Reich to vary the colour scheme 'according to the occasion, the mood or the weather'.[36]

Wood used curtains in a variety of ways: functionally, to obstruct light and draughts, for visual privacy; and for spatial and aesthetic effect. The importance that he placed upon them is reflected in the fact that the specification of works for Wildwood's ironmongery identifies tracks, runners and curtain hooks among those items that required the direct approval of the architect.[37] He often used curtains in preference to doors or partitions, which needed to be pinned back when opened, thereby reducing the space available for furniture and making small rooms feel somewhat smaller. At Wildwood, curtains were used to separate the main living area from the dining area and the kitchen, enhancing a sense of freedom of circulation around the ground floor. Where greater acoustic privacy was required for the occasional guest bedroom/playroom, folding timber panels were used. In other locations, where curtains were used for visual or spatial effect, tracks were concealed behind pelmets or between structural timbers. Those on the upper gallery allowed the visual separation of the sleeping accommodation to give privacy to guests at times when the living area was in use.

The selection of textiles was left to Wood's clients and Elisa Money chose the fabrics and made up the curtains herself.[38] Her choices included designs by Edinburgh Weavers, such as 'Full Measure' (1957), a cotton crepe designed by Kenneth Rowntree, which she used in the kitchen, and Hans Tisdall's screen-printed design 'Pheasant Moon' (1960), which she chose for one of the bedrooms.[39] The curtains for the living area required 40 yards of fabric to complete, a major undertaking for an amateur needlewoman.[40] In practice, these were only ever drawn at night, during the winter. Happy with her homemaking decisions, she lived for many years with these original textiles, relining curtains as they wore and only replacing them if the fabric was completely beyond saving.[41]

Wood's original colour palette for the living area was restricted to whites and muted tones. The painted brickwork of the fireplace wall was designed to complement the richer tones of the timber ceiling and walls, the cork flooring, and the large slate hearth that was a major feature of the interior.[42] His preference for natural materials can be contrasted with the style of interior decoration favoured by contemporary decorators, such as Noel Carrington. In a refusal of the intrusion of modern architects into the interior, Carrington argued, in his book *Colour and Pattern in the Home* (1954), that certain materials were unsuited to use within a domestic setting:

> The introduction into a room of materials normally used for the outside of buildings is hard to justify. The very qualities of brick or stone which make them able to endure the sun, frost and rain render them too hard and cold in the interior of a house. This is shown by their gradual disappearance in the history of decoration, apart from polished marble or other stones which could justify their position in the more palatial residences, and which are the natural building materials for Mediterranean countries. For some reason a vogue has developed for the intrusion of whole walls of granite, flint or red brick, not merely as the immediate surroundings of the hearth, but as the main decorative keynote of a living-room. It is commonplace amongst American architects, especially those setting out to be daring and wholly functional.[43]

Although the keynote of the interior was muted, Wildwood was not without colour. Red unglazed floor tiles brought warmth and practicality to the entrance area and the main wall of the dining room was covered in a striking hand-blocked wallpaper design – 'Pacific' from Sanderson – in a coral colourway that Wood suggested to his clients.[44] Brightly coloured tiles depicting birds, bought by the Moneys at Bentalls department store at Kingston, reflected the couple's personal interests and were fitted by Nigel Money in the downstairs cloakroom, one of several decorative additions that he made to his home.[45]

The kitchen was the principal element of the design of the house upon which Elisa Money was consulted, an acknowledgement of her authority in that area. Of significance, in light of the multiplicity of idealised images of modern kitchens that circulated in Britain in the 1950s, was the fact that her ideal kitchen derived from personal experience. Comfort was her primary concern and her main requirement, having experienced an unbearably hot kitchen in a former home, was for the kitchen of her new house to be cool.[46] Practical concerns, such as the need for early morning light and a suitable temperature in which to work had formed a focus for inter-war discussions about kitchen design, particularly in relation to the servantless, middle-class home. That emphasis became less marked as improved thermal efficiency permitted greater flexibility in the planning and organisation of the interior and as the wider availability and affordability of labour-saving devices began to make the conduct of heavy work less of an issue for many middle-class housewives. In its place a more social emphasis emerged in the writings of home economists and design reformers in response to the perceived isolation of women in the home, particularly within new post-war communities where family ties were often broken.[47]

The kitchen fittings for Wildwood were designed by Wood for completion by a joiner. At one end of the room a large fitment, incorporating a breakfast bar, separated it from the dining room. Service hatches between kitchens and dining areas had become a standard component of many interiors by the 1950s, but as H. Dalton Clifford observed, in his book *New Houses for Moderate Means*, these were becoming more elaborate in design. The advice that he gave to readers thinking of introducing such an opening into their own interiors centred on its labour-saving qualities:

> To be fully effective as a labour-saving device the hatch should be combined with double-sided china and glass cupboards and pull-through drawers for cutlery, glass, napkins etc. There should also be shelf space for storing pickles, jams, cereals, sugar, condiments and other non-perishable foods which do not have to be washed up or returned to the larder between meals.[48]

Kitchen fitments were central to the post-war transition of the modern, middle-class kitchen from a discrete, rationally conceived workspace to a more socially integrated environment. They not only brought counter dining, which had long been part of urban public dining experience, into a domestic context, but also helped to support new informal lifestyles, centred on leisure and home entertaining.[49]

The kitchen at Wildwood was designed to be large enough for casual dining, reflecting this expanded social vision for the space. In describing it *Country Life* magazine noted:

> To-day the cook is often the lady of the house, and the kitchen is regarded as part of the living space rather than as a workshop cut off from the reception rooms by close-fitting doors to keep out noise and smells.[50]

Woman magazine described it in social and spatial terms:

> Hub of the house is this counter fitment which divides kitchen from dining area. Here Elisa and Nigel have a snack, and talk over the day's doings. Everything is to hand, making it easy to provide tea and coffee for casual guests.[51]

Its design demonstrates Wood's attention to the details of domestic life in a very practical sense. From the dining side the unit was designed to conceal

Figure 4.7 Kenneth Wood. Wildwood, Oxshott. View from the kitchen to the dining
room, *c.* 1959.
Credit: Courtesy of Elisa Money.

the paraphernalia of the working kitchen from view – only crockery and
glasses could be seen on display – while to the kitchen side, open shelving
stored pots and pans, a cruet, instant coffee, marmite and a stainless steel
teapot (see Figure 4.7)

By the end of the 1950s increased prosperity had made open-plan living
an attainable aspiration for many young couples living in houses of conventional cellular plan.[52] As 'knocking through' became more popular, consumer magazines began to give greater consideration to the visibility of the
kitchen within the modernised interior. In May 1959, *Homemaker* magazine showed its readers a variety of possibilities for rearranging the interiors
of their older properties and points to an emerging emphasis on the aesthetic
modernity of the kitchen that was expressed in a more visual emphasis in
discussions of its design and planning:

> If you want to open up the plan of your older type small house, you can
> have the partition wall separating kitchen and dining-room removed

Figure 4.8 Kenneth Wood. Wildwood, Oxshott. Nigel Money with a scale model of the house, which he completed from Kenneth Wood's drawings, *c.* 1959.
Credit: Courtesy of Elisa Money.

and an open divider placed there instead. At the same time, you must modernize the kitchen equipment and decorate it so that it presents an attractive view of the dining room.[53]

Wildwood's kitchen was planned for convenient access to the front door, to receive guests and to take in the deliveries that remained a part of middle-class domestic life until supermarkets began to change shopping patterns.[54] Another consideration in locating the kitchen within the overall plan of the house was its relationship to the garden. *Ideal Home* magazine cautioned against allowing the kitchen to become a hazardous through route, particularly for children.[55] Many architect-designed houses of the 1950s attempted to overcome this problem by creating playrooms with separate access to the garden, as Wood did in his plan for Wildwood. Functional considerations informed Wood's approach to the design of the outdoor spaces of the house, particularly those adjoining the kitchen, which was connected to a utility room to the rear of the garage by means of a covered path. A herb

Figure 4.9 Kenneth Wood. Wildwood, Oxshott. Nigel Money at work landscaping the garden, *c.* 1959.
Credit: Courtesy of Elisa Money.

garden, which ran alongside it, was conveniently located for the back door and an open path to the rear of the garage allowed access for tradesmen. A sheltered side lawn, hidden from the front and rear of the house, provided a suitably unobtrusive location in which to dry laundry.

Wood's clients undertook a variety of activities that could be broadly understood as coming under the umbrella of do-it-yourself. These ranged from simple home repairs to creative homemaking activities such as the design of furniture and soft furnishings, the decoration of interiors and the landscaping and planting of gardens.[56] The additions and alterations that the Moneys made to Wildwood reveal the ways in which the spaces that Wood created for them were shaped in response to the pattern of their everyday lives. As the client for an architect-designed house, Nigel Money was an active agent in the creation of his new home and was involved in its design on many levels. His completion of a scale model of the house, using Wood's design drawings, suggests his enjoyment of the design process (see Figure 4.8). Elisa Money has indicated that her husband also played a significant role in the realisation of the interior, bringing his own views on design, colour and materials to bear on its furnishing and decoration.[57]

Over the years the Moneys supplemented Wood's interior fittings with additional elements that Nigel Money designed and completed. These included bookcases in the living area, a shoe rack in the hall, a knife rack in the kitchen, headboards and wall trims in the bedrooms and a dressing table in the guest bedroom. Through his creative and practical ability as a designer, Money's additions to the house extended Wood's original vision for the interior. Although many of the alterations and additions that Money made to his house might be understood as typical do-it-yourself activities, his approach was that of a professional designer.[58] This is perhaps best exemplified in his development of the exterior spaces of the house. Although Wood's design for Wildwood included plans for the exterior landscaping, this was not undertaken as part of the original building work but was left to his clients to complete.[59] Money, who was a serious amateur sportsman, gave up rugby to spend his weekends at the building site during construction. After the couple moved in he completed the hard landscaping to his own design and rather than turning to one of many contemporary sources of popular advice consulted Elizabeth Beazley's *Design and Detail of the Space Between Buildings* (1960) (see Figure 4.9).

Wildwood incorporated what were to become essential elements of Wood's domestic architecture: a sensitive approach to site and planning; the use of structural timber to allow a freer approach to planning; the creation of borrowed views to enhance a sense of interior spaciousness; the conjunction of fixed and movable elements – built-in furniture, curtains and partitions – for social and spatial flexibility; the use of natural materials for their visual and symbolic values; attention to economy and the need for ongoing maintenance; the generous provision of services; and a strong commitment to meeting the personal requirements and aspirations of his clients.

Notes

1 In their study of flexible housing, Tatjana Schneider and Jeremy Till use the terms 'soft' and 'hard' to describe the way in which space is planned and used: 'Soft use allows the user to adapt the plan according to their needs, the designer effectively working in the background. With hard use, the designer works in the foreground, determining how spaces can be used over time'. These two approaches, they suggest, evolved concurrently within architectural modernism, in the open plan and in the modernist concept of minimal dwelling, the former allowing users to establish the use of space and the latter, 'more architect-determined' expressed in folding and unfolding interior elements that could be altered in response to different needs. See, Tatjana Schneider and Jeremy Till, *Flexible Housing* (London: Architectural Press, 2007), 7, 19.
2 Elisa Money, in discussion with the author, October 2011.
3 Elisa Money, in discussion with the author, October 2011.

4 See, Kenneth Powell, *Powell & Moya* (London: RIBA, 2009), 23–5. See also, 'England. Two houses at Oxshott, Surrey. Powell and Moya', *AD* (March 1956): 96.

5 Mrs Money recalls them as the only houses of interest that they saw. Elisa Money, in discussion with the author, October 2011.

6 Elisa Money, in discussion with the author, October 2011 and EMPP.

7 Elisa Money, in discussion with the author, October 2011.

8 Elisa Money, in discussion with the author, October 2011.

9 Elisa Money, in discussion with the author, October 2011.

10 Elisa Money, in discussion with the author, October 2011.

11 Elisa Money, in discussion with the author, October 2011.

12 Elisa Money, in discussion with the author, October 2011.

13 Houses built in the area in the 1950s were modest in comparison to those being built on the same estate today. A new build near to Wildwood, marketed in 2013, provided 7,500 square feet of accommodation on 0.75 of an acre in comparison to Wildwood's 1,500 square feet on a slightly smaller 0.705 of an acre.

14 The galleried form was acknowledged as suitable for the smaller house. Raymond McGrath noted that although it was not 'the most private system' it gave 'a good feeling of space'. See, McGrath, *Twentieth Century Houses*, 33, 91.

15 On timber construction and the use of weatherboarding in Surrey see, John and Jane Penoyre, *Houses in the Landscape: A regional study of vernacular buildings styles in England and Wales* (London: Faber & Faber, 1978), 23–33 and Roderick Gradidge, *The Surrey Style* (Kingston: The Surrey Historic Buildings Trust, 1991).

16 The influence of building restrictions stimulated the development of more open living spaces from the early the 1950s. One of the earliest British examples of a house with a modern open plan, built while restrictions on total floor size remained in place, is 22 Avenue Road Leicester, commissioned by Mr and Mrs Goddard in 1953 and designed by Fello Atkinson and Brenda Walker of James Cubitt and Partners. See, Harwood, *England: A Guide to Post-War Listed Buildings*, 2nd rev. edn. (London: Batsford, 2003)136.

17 An notable source on Wright, with which Wood was familiar, was Sigfried Giedion's *Space, Time and Architecture* in which Giedion devotes a section to Wright's influence in Europe. See Sigfried Giedion, *Space, Time and Architecture: The Growth of a New Tradition* (Cambridge: The Harvard University Press, 1942), 319–48.

18 Trevor Dannatt has recalled that his third year theory lessons, taught by Emil Scherrer, were based on Giedion's book. Trevor Dannatt, interviewed by Alan Powers, 2001, tape F11643, Architects' Lives Series, British Library.

19 Giedion, *Space, Time and Architecture*.

20 On Wright's use of multiple internal views see, Sandy Isenstadt, *The Modern American House: Spaciousness and Middle Class Identity* (Cambridge: Cambridge Unversity Press, 2006), 67. Wood's plan for Tanglewood was included in 'Space saving and space', a supplement to *Homes and Gardens* magazine in November 1962. The interior had no doors between the main living areas and used changes in floor materials to articulate nominal thresholds between spaces. The magazine commented on Wood's incorporation of 'borrowed' views to create an awareness of 'more space beyond in the glimpse of the extending floor and the hint of further furniture grouping.' See 'Space saving and space', 12.

21 H. Dalton Clifford, *New Houses for Moderate Means*, (London: Country Life, 1957), 24.

22 Charlotte Benton, 'Le Corbusier: furniture and the interior', *Journal of Design History* 3, 2/3 (1990): 103–24, on 110.

23 Wells Coates from the 24 May 1933 broadcast, 'Modern Dwellings for Modern Needs', in *Documents: A Collection of Source Material on the Modern Movement* (Milton Keynes: The Open University Press, 1979), 73.

24 As Pat Kirkham has indicated, such indoor/outdoor lifestyles were not always viable, even in the warmer climate of California. Pat Kirkham, 'At home with California Modern, 1945–65', in Wendy Kaplan, ed. *Living in a Modern Way: California Design, 1930–65* (Cambridge, MA: MIT Press/Los Angeles County Museum of Art, 2011), 146–76.

25 This can be seen in Wood's work and is also evident when comparing architects' own homes to those that they designed for private clients. Peter Aldington is a case in point. His house at Haddenham incorporated fixed seating and a concrete platform bed that expressed a clear vision for the space based on permanent social and functional needs. It can be contrasted with the interiors that he realised for clients, such as those for Diggs Field at Haddenham (1967–69), which were designed to accommodate loose items of antique and modern furniture. See, Alan Powers, *Aldington, Craig and Collinge* (London: RIBA, 2009), 23–38.

26 This is most notable in the designs for the Picker House interior and the zoning of the main living space.

27 See, for example, *Country Life* (1 October 1959); *Wood* (August 1959); *A&B* (September 1959); *Deutsche Bauzeitung* (March 1960); *House and Garden* (September 1959, December 1959 and November 1960); *Woman* (21 May 1960); *Financial Times* (13 November 1961).

28 Elisa Money, in discussion with the author, October 2011.

29 Edith Blair, 'House of today', *Woman* (21 May1960): 13.

30 Blair, 'House of today', 13.

31 Open storage units became popular in the 1950s and lent themselves to all manner of domestic display. As Penny Sparke has indicated, such domestic displays undermined the rationalising impulse that had informed their introduction. See, Penny Sparke, *As Long as it's Pink: The Sexual Politics of Taste* (Halifax, Nova Scotia: The Press of the Nova Scotia College of Art and Design, 2010), 128.

32 Elisa Money, in discussion with the author, October 2011.

33 Elisa Money, in discussion with the author, October 2011.

34 Christine Boydell, 'Textiles in the modern house', *The Journal of the Twentieth Century Society* 2, 'The Modern House Revisited' (1996): 52–64, on 54. See also Joel Sanders, 'Curtain wars: architects, decorators, and the 20th-century domestic interior', *Harvard Design Magazine* 16 (Winter/Spring 2002): 14–20.

35 H. Dalton Clifford, *Country Life Book of Houses for Today* (London: Country Life Limited, 1963), 44.

36 Clifford, *Country Life Book of Houses for Today*, 66.

37 Specification of Works, EMPP.

38 Elisa Money, in discussion with the author, October 2011.

39 The design was available from Heals, Liberty and Woollands from September 1961, suggesting that the couple may have replaced their original curtains, or managed without in their early years of occupation. 'Fabrics for the walls', *The Times* (5 June 1961): 17.

40 Elisa Money, in discussion with the author, October 2011.

41 Elisa Money, in discussion with the author, October 2011.

42 Wood acknowledges Arne Jacobsen as an influence on his work at this time. As Alan Powers has indicated, white painted brick had become popular in Denmark, perhaps in response to Jacobsen's Klampenborg houses (1946–51). See, Powers, *Aldington, Craig and Collinge*, 21.

43 Carrington, *Colour and Pattern in the Home*, 140.

44 Elisa Money, in discussion with the author, October 2011.

45 Elisa Money, in discussion with the author, October 2011.

46 Elisa Money, in discussion with the author, October 2011.

47 Within the context of mass housing this led to new approaches to planning, including the orientation of kitchens to the front of the house. See, Judy Attfield, 'Inside Pram Town: a case study of Harlow house interiors, 1951–61'. In *Bringing Modernity Home: Writings on Popular Design and Material Culture* (Manchester: Manchester University Press, 2007).

48 Clifford, *New Houses for Moderate Means*, 23.

49 As Grace Lees-Maffei has shown, these new environments and the forms of domestic advice that interpreted them for a contemporary audience also helped to shape new forms of gendered social performance within the home. See, Grace Lees-Maffei, 'Dressing the part(y): 1950s domestic advice books and the studied performance of informal domesticity in the UK and the US', in Fiona Fisher et al., eds. *Peformance, Fashion and the Modern Interior: From the Victorians to Today* (Oxford: Berg, 2011), 183–96.

50 H. Dalton Clifford, 'Designed for the modern way of life', *Country Life* (1 October 1959): 426.

51 Blair, 'House of today',13.

52 Terence and Shirley Conran were among those young couples pioneering these new urban lifestyles in the late 1950s. In 1957 they described their experience of homemaking for *House and Garden* magazine. The magazine featured photographs of their Regency terraced house, one floor of which had been knocked through from four rooms to create a large living/dining/kitchen space. 'The Terence Conrans talk about setting up house the hard way', *House and Garden* (June 1957): 46–9.

53 Edgar Lucas, AIAA. 'Conversion to open plan', *Homemaker* (May 1959): 304–5, on 304.

54 A useful and practical feature that Wood included in a number of designs was an external trade hatch or cupboard, recessed in an outside wall – a feature first introduced to the public through the *Daily Mail*'s Ideal Labour-Saving Home competition of 1920. On the trade hatch designed for the Ideal Labour-Saving Home competition see Ryan, *The Ideal Home*, 34.

55 In November 1960 *Ideal Home* magazine published a kitchen supplement incorporating recommendations from its kitchens commission, which it hoped would inform consumers and influence the designs of manufacturers. The members were: Alan Gore, Jeremy Dodd, Elizabeth Gundrey, Colin MacGregor, Mrs Ronald Searle and Mrs Christopher Robinson, with the *Ideal Home* staff architect and domestic science experts. The model kitchen that they envisaged was a space in which the spatial separation of kitchen, pantry, scullery, and laundry could be retained in compressed version within the modern plan. See, 'Your guide to a better kitchen', *Ideal Home* (November 1960), supplement.

56 Paul Atkinson has suggested that do-it-yourself activities can be characterised according to motivation, ranging from creative activities to forms of maintenance and lifestyle-oriented improvements. Paul Atkinson, 'Do it yourself: democracy and design', *Journal of Design History* 10/1 (2006): 1–10, on 3.

57 Elisa Money, in discussion with the author, October 2011.

58 Television programmes, such as those presented by Barry Bucknell from the mid-1950s, represented do-it-yourself as a legitimate sphere of masculine engagement with the home. The garden was another context in which men were active participants in the development of their domestic environments. See Andrew Jackson, 'Labour as leisure: the Mirror dinghy and DIY sailors', *Journal of Design History* 19/1 (2006): 57–67; Stephen Constantine, 'Amateur gardening and popular recreation in 19th and 20th centuries', *Journal of Social History* 14/3 (1981): 387–406.

59 His original drawings show a curving driveway to the front of the house and two terraces for relaxation, one adjoining the playroom/guest room and another to the rear of the house, incorporating seating and a sculpture pool. A sculpture position was also indicated to the front of the house. Neither setting was realised, which perhaps suggests that they were more a reflection of Wood's wider artistic interests than a priority for his clients.

5 Timber house

British Columbia Lumber Manufacturers' competition house, 1957

In 1957 the British Columbia Lumber Manufacturers' Association (BCLMA) launched an architectural competition to promote the use of Canadian forest products in British domestic architecture. Wood had already begun to pursue an interest in timber construction through his early church work and his first two private house commissions. His successful entry to the competition gave him the opportunity to travel to North America and study the work of the lumber industries at first hand, and he extended the trip to visit a number of important buildings by leading modern architects in Canada and the United States. His involvement in the competition was significant for the development of his practice and offers an opportunity to consider the wider cultural impact of the activities of the Canadian lumber industries on the design of British houses of the 1950s and 1960s.

Although timber houses appeared in a number of inter-war publications on modern architecture, timber was not used greatly as a structural material in Britain at that time. The impact of Serge Chermayeff's house, Bentley Wood (1938) (see Figure 5.1) – the most significant British example of a modern timber house of that period – was not felt fully until after the war, when John Summerson described it as 'easily the most memorable' of the country houses built in England in the 1930s and 'the most aristocratic English building of the decade'.[1] Through his use of materials and reconsideration of the relationship between modern art, architecture and landscape, Chermayeff pointed a way forward that was taken up by Wood and others in the 1950s as they looked for ways to balance technological advances with a more human-centred and contextual approach to architectural design.

Research into the use of timber and new timber products in British housing gathered pace from the 1930s as new materials began to challenge its dominance in some areas of building. In 1934 the Timber Development Association was established to 'mount a technical challenge to the increasing threat of steel windows in a market traditionally dominated by timber'.[2] Two years later it ran a timber housing competition. The winning

Figure 5.1 Serge Chermayeff. House at Bentley Wood, Halland, East Sussex, 1938. Credit: Dell & Wainwright/RIBA Library Photographs Collection.

design – a pitch-roofed, timber-framed house – was exhibited at the Ideal Home Exhibition at Olympia in 1936.

Designers also began to give greater consideration to the structural use of timber in modern architectural design, among them Raymond McGrath and Robert Furneaux Jordan, who began to position it within a modernist architectural frame and in relation to new materials. Raymond McGrath believed timber to be most appropriate 'for the country or week-end house' and included several examples of timber buildings in *Twentieth-Century Houses*, among them a summer residence at Karuizawa in Japan by Antonin Raymond.[3] Dividing contemporary houses into those of 'solid wall' construction and those of 'frame structure with curtain walls' he drew parallels between steel and timber-framed buildings, commenting on Rudolph Schindler's Lovell House in California:

> The steel frame is open to the view, and its regular parallel supports give the house the look of a delicate Japanese structure, for steel and wood are much nearer to one another than they are to brick and stone.[4]

Robert Furneaux Jordan made similar observations in an article for *The Times* newspaper in 1936:

> One of the most impressive facts about timber on the structural side is its adaptability to modern methods of planning. The wide spans required by the horizontal windows and flexible planning of the modern house have been the natural results of steel and concrete, but they might equally well have been evolved for a timber style of building.[5]

These discussions of the mid 1930s helped inform a more open outlook to the use of materials in British work of the second half of that decade. Even so, building practices were slow to change and in the 1951 edition of *The Modern House*, F.R.S. Yorke noted that English construction was 'usually solid brickwork with steel or concrete girders over the wide openings, with either exposed or rendered brick surfaces, or in rarer cases ... monolithic concrete'.[6] Architects were important advocates for the wider use of timber in the 1950s and Wood was among those who wrote on the subject for a professional readership. Architects also helped to shape popular opinion through the contributions to popular magazines. In an article on timber for *Home* magazine, Michael Manser argued:

> These days there are so many good reasons for building houses in timber that it is hard to decide which to put first. But the most compelling for me, personally, is an inhibition about bricks and mortar which, in 1961, I consider to be a laborious, messy, disagreeable, inaccurate and generally outmoded way of building a house.[7]

The lumber manufacturing nations, Canada in particular, engaged in a variety of activities to promote the use of timber among British architects and inform British consumers of the benefits of timber houses. At the beginning of the twentieth century, British Columbia's forest industries began to cooperate with the objective of developing an international market for wood products.[8] Between 1918 and 1929 exports of British Columbia timber rose from 5 to over 27 per cent. Britain was the largest importer for much of that period, but was overtaken by Japan by the end of the 1920s.[9] Canada's timber building tradition made the design of the home a natural arena for consumer-oriented activities. Canada exhibited a bungalow made entirely from British Columbia timber at the British Empire exhibition in London (1924–25). While the house stimulated public interest it was not successful in generating sales.[10] In the face of the collapse of the American market for Canadian timber during the Great Depression, international marketing efforts were increased and in the 1930s the British Columbia Lumber Manufacturers' Association established permanent trade extension offices in Britain, South Africa and Australia.[11] British demand for Canadian timber, which increased during the Second World War, continued with the award of large two-year contracts to support reconstruction.[12] Following this period of high demand, competition from Sweden, Finland and the USSR began to affect trade. The BCLMA stepped up their attempts to develop the British market in the mid 1950s, advertising a special timber exhibit at the Building Exhibition at Olympia in 1955.[13] Subsequent national initiatives to promote the Canadian lumber trades followed the establishment of the Canadian Wood Development Council in 1959. These included the launch of National Forest Products Week in 1960, a major government-backed initiative to promote the lumber trades in Canada and the United States.[14]

Another way in which the Canadian lumber industries promoted timber products in the 1950s was through the sponsorship of architectural competitions. Canada's Trend House program (1952–54) and its less well-known British counterpart, the BCLMA competition (1957) were two such initiatives.[15] The Canadian programme was launched by a collective of lumber-related industries in British Columbia.[16] Designed by leading Canadian architects, eleven Trend Houses were built between 1952 and 1954: two in Toronto and one each in Victoria, Vancouver, Edmonton, Calgary, Halifax, Winnipeg, Regina, London and Montreal. The houses made extensive use of western woods – Pacific Coast hemlock, Western red cedar and Douglas fir – structurally, and within their interiors.

The first Trend House at Thorncrest Village, a Toronto suburb, opened to the public from May to December 1952 and showcased the best contemporary Canadian design.[17] It incorporated modern furnishings from Eaton's, Canada's leading department store, and included items from the newly established National Industrial Design Council's Design Index, which had been set up to promote Canadian furniture and industrially designed products to consumers.[18] The house, which was later sold to a private buyer, was aimed at the small family of average income and provided 1,000 square feet of living space, arranged in an open-plan layout. Subsequent Trend Houses followed the same model of public exhibition and sale and were built in the winter and spring of 1953–54 and advertised to consumers in the spring and summer of 1954.[19] The Trend Houses helped to shape public attitudes to modern design and stimulated interest in a new type of house that was easy to maintain, featured open-plan interiors, and was designed with concern for the natural landscape.[20]

A Canada Trend House was shown at London's Ideal Home Exhibition in 1957, where it appeared alongside show houses of more conventional appearance, such as the Crouch Convertible House by Kingston housebuilders, G.T. Crouch, which conformed to 'traditional design' but incorporated 'modern' features (see Figure 5.2).[21] Designed by Wells and Hickman, the Canada Trend House was furnished by Heal's department store and sponsored by the BCLMA in cooperation with the Canadian government. Promoted to the public as 'a Canadian timber frame house designed for the British Isles' it aimed to demonstrate 'the adaption of a modern building system – used for the vast majority of North American homes – to aesthetic and practical needs in this country'.[22] The Ideal Home Exhibition catalogue in which it appeared communicated a series of promotional messages

Figure 5.2 Wells and Hickman. The Canada Trend House, Ideal Home Exhibition, Olympia, London. The living room with the dining room beyond, 1957.
Photographer: John Maltby. Credit: John Maltby/RIBA Library Photographs Collection.

about timber homes to consumers and the building trade. These addressed practical concerns as well as the potential that timber houses offered for social differentiation, key messages being: 'frame houses can be built quickly at low cost', 'frame houses are warmer, more comfortable, save fuel' and 'frame houses give wider scope for individual design'.[23] The Canada Trend House also featured in a British Pathé news item in which a female demonstrator performed a leisured, middle-class lifestyle within its interiors – a significant means through which the spaces and technologies of the modern home were mediated in this period.[24] Astragal commented on the Ideal Home Exhibition houses in the *Architects' Journal*:

> The spec built houses are much as one would expect, but the timber Canadian house is worth a careful look, though I have my doubts about the heating system being really adequate. The interior (by Heals) has almost the only interesting furniture in the show.[25]

In the same year, the BCLMA launched a British architectural competition. Loosely based on the Canadian Trend House initiative, it aimed to develop a set of original timber house designs for promotion to British buyers. Ten

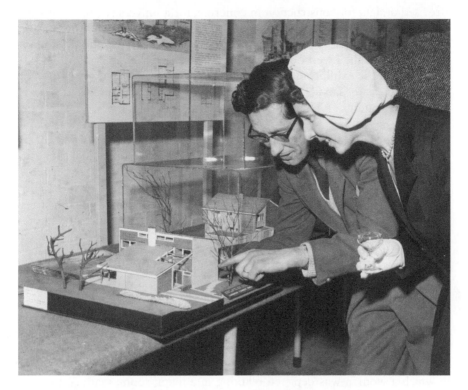

Figure 5.3 Kenneth and Micki Wood in front of a model of Wood's BCLMA compe-
            tition design at the Building Centre in London, 1958.
Credit: Courtesy of Kenneth Wood.

British architects from around the country were selected from over 300 who
submitted competition designs. Wood was invited to join the architects' panel
along with Leslie and Peter Barefoot (Suffolk), Edward Butcher (Dorset),
Munce and Kennedy (Belfast), C. Wycliffe Noble (Surrey), James Houston
and Son (Ayrshire), Bartlett and Gray (Nottingham), Philip R. Middleton
and Partner (Middlesborough), Robert Paine and Partners (Canterbury) and
Nelson and Parker (Liverpool). Their winning entries, which drew to varying
degrees on British and Canadian domestic prototypes, were produced for
display as models rather than fully realised show houses and were exhibited at the Building Centre in London from 26 March to 3 April 1958 (see
Figure 5.3). Architectural plans for the winning designs were made available to the public on a £25 royalty basis through the Timber Development
Association, which had offices in London, Bristol, Manchester, Birmingham,
Leeds and Glasgow. The total number of houses that were eventually built as
a result of the competition is unclear. Wood's design was completed at least
once, for a company director at Saffron Walden in Essex. He also produced

a revised design, for construction in concrete, for a university professor who planned to build it in Sierra Leone.[26] Nelson and Parker's exhibition house was also built at least once and formed a focus for a symposium on 'Modern Homes' held at the Liverpool Architectural Society in February 1959.[27]

Models of the exhibition houses featured in a British Pathé news item of 1958 to illustrate Canada's 'influence on the design of the homes of tomorrow in this country'.[28] The voiceover noted:

> The models bear little resemblance to the old log cabin that a lot of us associate with wooden houses, but besides looking attractive they can be put up in half the time it takes for the usual house and are certainly more economical to build. Other advantages, it's claimed, include more than twice the insulation of brick cavity walls, making central heating easier and cheaper. Here, at last, is a refreshingly new idea for home-building. Let's welcome the innovation.[29]

Selected plans and drawings of the competition houses were published in British architectural journals and each of the designs featured in a *Housing Trends* brochure that was available free of charge from Canada House in London (see Figure 5.4).[30] Very similar in editorial content to the promotional brochure that had been produced for the Canadian Trend House programme, *Housing Trends* incorporated a consistent set of marketing messages for the lumber industries. Both brochures contained a page entitled 'Gracious Living' in which a woman was depicted relaxing in a living room, reading a magazine (see Figures 5.5 and 5.6). Wells and Hickman's Canada Trend House provided the model for the British example of gracious living, which, when viewed alongside its Canadian counterpart, suggests differences between the two national contexts. In the British interior timber is incorporated within a broader palette of colours and materials, reflecting the contemporary style of design and decoration that came to prominence with the Festival of Britain. In contrast, the dominance of timber in the Canadian interior perhaps suggests greater confidence in its use and appeal to a domestic audience.

The text of the British *Housing Trends* brochure reveals a clear aspiration to attract the interest of speculative developers as well as that of individual consumers. Several of the commentaries proposed the suitability of the designs for series construction and mentioned the possibility of introducing variety to groupings through alterations to the cladding styles. In the case of Design 103, by C. Wycliffe Noble, whose practice was based at Sunbury on Thames, quite close to Wood's, the supporting text stressed the suitability of timber for English homes:

> Use of timber frame construction with brick facings and white painted shiplap or Western red cedar gable cladding, which weathers to silver grey not only gives the house a high standard of insulation but imparts a delightful English character to a design embodying the best techniques of economic building.[31]

*Figure 5.4 Housing Trends. A Selection of Designs Prepared by British Architects
Featuring Canadian Timber Frame Construction.*
Credit: Courtesy of Kenneth Wood.

Wood's competition house, Design 109, provided 1,380 square feet of living space at a cost of between £4,450 and £4,750 (see Figure 5.6). It was planned with a highly glazed living area, split into upper and lower zones: an upper living zone and adjoining terrace and a lower kitchen/dining zone that could be sub-divided by means of a folding timber partition to create a playroom with its own play terrace.[32] A dual-aspect fireplace served both levels. Also at entrance level were a guest bedroom, or maid's room, a cloakroom and a utility room. The first floor contained three further bedrooms, a bathroom and a separate w.c. Design 109 was planned for expansion. The garage was sited to allow its future incorporation into the main living area and a further garden terrace, to the rear, was envisaged as another potential site for extension. Wood's was one of two competition houses – the other by Philip Middleton and Partner – designed for construction on either a flat or a sloping site, employing what *Housing Trends* described as 'the North American principle of split-level planning'.[33]

In April 1958 *The Lady* reviewed the Canadian National Exhibition in Toronto and included in the same issue an article on the British competition

Gracious Living

The warm, appealing beauty of Canadian timber creates an atmosphere of friendliness that transforms a house into a truly gracious home. Wood finishes possess outstanding durability and they are easy and economical to maintain. This charming room, with its attractive, cedar-panelled fireplace wall, reveals some of the countless imagination effects that can be achieved with Canadian timbers from British Columbia. Further details of the house, which was built and furnished by Heal's, are given on the inside back cover of this booklet.

Figure 5.5 Gracious living UK style. From: *Housing Trends*.
Credit: Courtesy of Kenneth Wood.

houses, 'Built in Canadian Style'. The Canadian National Exhibition was not the only pretext for the article, which also made reference to the presentation to Parliament of the Thermal Insulation (Dwellings) Bill, which it understood as an incentive to timber construction in the United Kingdom.[34] The magazine, which illustrated two of the competition designs, a small bungalow by Bartlett and Gray of Nottingham (£2,900–£3,150 excluding land) and Wood's design for a larger family home, advised readers of their suitability for both urban and rural locations:

> All the designs are ambitious and imaginative and yet are far from beyond the reach of the average income, at prices from £1,750 up to £4,850, exclusive of the cost of the land. In each design – whether for a retired couple or for a young household where extra rooms may be needed later on – timber has been used both for the actual framework of the house and for all or part of the external cladding. Brick and wood, or stone and wood, combine very pleasingly and, used together, enable an architect to design a timber frame house on a site which might otherwise be unsuitable. Under the Model Bye-Laws, timber clad walls must

The warm, appealing beauty of Western Woods creates an atmosphere of charm and friendliness that transforms a house into a truly gracious home. And the outstanding durability and easy upkeep of Western Woods offer you lifetime economy of maintenance. This lovely room, with its handsome, panelled fireplace wall and beamed ceiling reveals some of the countless imaginative effects you can achieve easily and economically when you build or remodel with Western Woods.

Figure 5.6 Gracious living Canadian style. From: *Western Woods Present Ten Canadian Trend Houses.*
Credit: Courtesy of Michael Kurtz.

be at least 10 feet from the boundary of the plot in a built-up area; but where brick or stone is used, or where timber frame walls are rendered, the distance is reduced to 5 feet.[35]

As the author indicated, certain aspects of their design, principally their spatial organisation and use of technology, marked the competition houses out from more conventional British designs:

> Spatial effects are achieved on relatively small plots of land. A car port in place of a garage, and the grouping of rooms to enclose little inner courts or gardens, give variation to the pattern of house design normal to this country, and the ease with which ducted warm-air heating can be run through the hollow walls makes possible a much more comfortable and economically-run home.[36]

The article attempted to dispel concerns about the cost of fire insurance and the availability of mortgages and discussed the savings that could be made

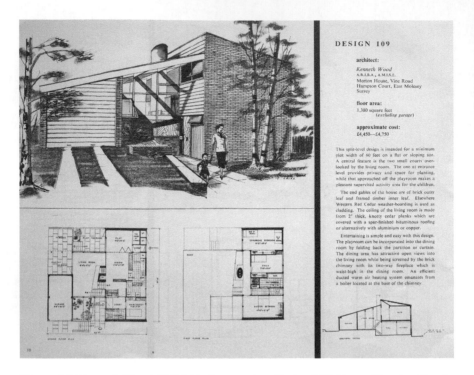

Figure 5.7 Kenneth Wood, BCLMA Competition, Design 109. From: *Housing Trends.*
Credit: Courtesy of Kenneth Wood.

using dry construction – plasterboard for interior walls instead of wet plaster, allowing the rapid installation of services without costly delays due to bad weather. Although the competition appears to have generated interest, British lenders remained reluctant to offer mortgages for timber houses at a time when modern designs of all types could be difficult to finance. Buyers of one of Span's brick cross-wall construction terraced houses at Blackheath (1957) had their mortgage application 'repeatedly turned down because the house was modern' and felt that building societies were 'the main enemies of modern architecture' at that time, a sentiment echoed in other contemporary sources.[37] In the face of such difficulties, the BCLMA arranged a British lecture tour by the Canadian mortgage authority in 1959.[38]

Combined educational and promotional efforts, of which the British architectural competition represents one example, focused on a number of core themes to change public perceptions in domestic and overseas markets. Through them, an emerging set of modern domestic values began to coalesce around materials in the late 1950s. In the United States, the National Lumber Manufacturers' Association produced a free twenty-page colour booklet entitled *Livability Unlimited: With Homes and Products of*

Wood. The booklet was publicised in a major advertising campaign that featured in magazines such as *Life* from the late 1950s.[39] The phrase 'livability unlimited' was also adopted in Canada, where wood was promoted for its emotional appeal, warmth and beauty, durability and economy, insulation, acoustic properties, and flexibility in permitting the expansion of buildings without the need for expensive structural changes.[40] Flexible expansion was represented as 'an advantage in family dwellings and a particularly important feature of public buildings, where the rapid growth of a community can make a structure obsolescent in a very few years' – facts with which Wood was well-acquainted through his own experience of working in Surrey's expanding suburbs.[41]

A consistent approach to the promotion of timber can also be seen the Canadian lumber industry's professionally focused activities. In Britain, advisory groups were set up from the mid 1950s with the aim of changing perceptions within the architectural profession and the building trades. The Plywood Manufacturers' Association of British Columbia opened an office in London in 1956 and appointed a technical field representative to give advice on Canadian fir plywood to British architects, engineers and contractors.[42] In 1964 the British Columbia Timber Frame Housing Group was created as a permanent advisory group to British architects, contractors and local housing authorities to gain wider acceptance for timber frame housing.[43] In the same year, John Laing Construction Limited built three pairs of timber houses at Abbots Langley in Hertfordshire for the Canadian Department of Trade and Commerce to promote a platform frame system of construction.[44]

Promotional brochures such as *Trends in Timber Construction, Frame Construction with Canadian Timber* and *Wood-Frame House Construction* were distributed through Canada House. *Frame Construction with Canadian Timber* featured two of Wood's designs, Wildwood (Chapter 4) and Fenwycks (Chapter 7). The publication outlined the merits of the material: 'speedy construction', 'warmth and comfort with economy', 'immediate occupation', 'frost-free plumbing', 'scope for individuality', 'appeal to builders', 'prefabrication possibilities' and 'performance'.[45] A single page advertising campaign in British architectural journals had a similar educational focus, each centring on a type of Canadian wood (Canadian Douglas fir, spruce, red pine, white pine, Western red cedar and Pacific coast hemlock) and its practical applications and advantages.[46] The cultural currency of Canadian home and its associated values of warmth and comfort was exploited in other contexts. Contemporary advertisements by Pilkington Brothers, for example, used the image of a couple in a Canadian home, looking out at a winter landscape, to promote Insulight double glazing.[47]

The Canadian home remained a reference point for architects and consumers in the early 1960s, with Canadian style houses exhibited at the Ideal Home exhibitions of 1960 and 1961. Advertised by its builders, John Maclean and Sons of Wolverhampton as 'the new Canadian Trend House', the Maclean Vancouver House of 1961 was furnished and decorated by *New*

Homes Magazine and featured 'large sliding picture windows of Canadian style' along with 'generous built-in wardrobes' and a 'separate laundrette/mudroom'.[48]

The British participants in the British Columbia Lumber Manufacturers' Association competition were offered a fee for their participation as architectural panellists or the option of a return air ticket to Canada to study the work of the lumber industries at first hand. Wood took advantage of the offer of a transatlantic air ticket and made a whistle-stop architectural tour with another participant, Peter Barefoot. The trip took place in May 1958, and Wood recorded his impressions in photographs and writings that formed the basis for several articles on Canadian architecture and building techniques that he published and broadcast on his return.[49] They began their trip in London, flying to Amsterdam and on to Vancouver via Sondrestrom. Flying over Canada, he observed little sign of development until the outskirts of Edmonton, which appeared from the air like a 'vast caravan site'.[50] In Vancouver, a flourishing centre of modern design, Wood observed 'the first signs of a really indigenous art' but was also struck by the almost unchecked development of the city.[51] In his travel journal, he recorded his thoughts on the decaying inner suburbs, the 'tyranny of the automobile' and the rapid suburbanisation of Vancouver's North Shore.[52] Along with general comments on Canadian culture, it is his experience of the speed of urban change that leaves a lasting impression on the reader – of the newly built highways with their 'ragged verges' and of a 'bulldozer mentality' that appalled him in its 'wholesale stripping of housing development areas' and destruction of the natural landscape.[53] The staccato style of Wood's prose evokes Ian Nairn's polemical writings on the British suburbs of two years earlier, suggesting that these were still very much in his mind.

Wood was a critical observer of Canadian building technique, noting that the workmanship and external finishes of houses were generally rougher than at home and that the detailing was poor 'even in the best private houses, including architect owned'.[54] He made notes on platform and post-and-beam construction, observing its use in local housing design and in public buildings such as churches and schools.[55] His sense of the local architecture was that it was often imaginative, an 'easy familiarity' with timber products resulting in houses that were not only comfortable and efficient but also 'honestly modern in approach'.[56] He found much to admire and 'surprisingly little sign of any logcabin cozy sentimentality'.[57] And he saw indications of affluence and a new attitude to the home, centred on consumption – 'Gadget homes, H.P., car whatever cost' – that was increasingly evident in British contemporary culture.[58]

His engineering services training made him particularly attentive to the quality of services incorporated in the buildings that he encountered. The local practice of using sloping sites to accommodate basements for heating, storage and laundry interested him, but he observed, somewhat disparagingly, that split-level planning was also 'applied to flat sites as fashion

Figure 5.8 Wood's photograph of the John Porter House, West Vancouver, 1947–48, taken on his visit to Canada in 1958.

Credit: Courtesy of Kenneth Wood.

rather than logic'.[59] He found limited evidence of prefabricated construction at the sites that he visited, due to the ready supply of surfaced timber and felt that 'more mechanisation, large scale prefabrication and ready availability of standard sizes' would be necessary if timber construction was to become economically viable in Britain.[60]

During his stay he visited buildings by leading modernist architects, among them William Wilding's Highland United Church, the Forest Products Research Laboratory at the University of British Columbia by Berwick Thompson and Pratt and St Anselm's Church by Semmens and Simpson. He also visited examples of modern speculative housing by the Lewis Construction Company as well as architect-designed homes, the most significant of which was John Porter's Massey award-winning home of 1947–48. The Porter House was of a locally distinctive style of modern architecture that became nationally influential in the 1950s through its dissemination in magazines such as *Western Homes and Living* and *Canadian Homes and Gardens*.[61] Most were built as family homes in suburban areas of North or West Vancouver. Uneven ground made the use of traditional building techniques difficult and stimulated the development of post-and-beam construction.[62] Porter used a rock outcrop to one side of the plot to introduce a change in level within the interior of the house, which was

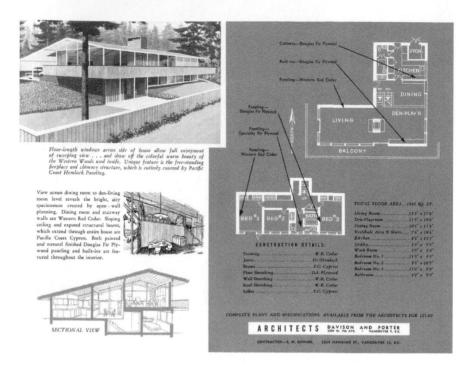

Figure 5.9 Davison and Porter, Vancouver Trend House. From: *Western Woods Present Ten Canadian Trend Houses.*
Credit: Courtesy of Michael Kurtz.

split into an upper living room and a lower play area, divided by a natural planted bank.[63] Basil Spence, who visited the house three years before Wood, described it as a 'butterfly settled on a leaf'.[64] Wood's informal snapshot shows the geometry the design and the interplay of light on its surfaces (see Figure 5.8). One of Davison and Porter's designs, a split-level house with open wall planning, built at Skyline Drive in North Vancouver in 1954, had formed part of the original Canada Trend House programme (see Figure 5.9).

In Toronto Wood found much of the new architecture 'disappointing' but was impressed by the work of John Parkin Associates at Don Mills.[65] Many of his notes and comments found their way into finished articles, such as 'Timber Construction in Canada' which was published in *The Architects' Journal* in October 1958.[66] Although Wood's diary records little of his time in the United States, his notes and photographs show that he visited houses by Frank Lloyd Wright at Oak Park; Charles Beersman's Wrigley Building (1920–24) with its new plaza (1957); Skidmore, Owings and Merrill's Inland Steel Building (1958), which was Chicago's first commercial high-rise building of the post-war period; and examples of Mies van de Rohe's work at Illinois

Figure 5.10 Kenneth Wood, Forestry Commission District Office, 1968.
Photographer: Colin Westwood. Credit: Colin Westwood/RIBA Photographs Collection.

Institute of Technology, including Crown Hall (1956), as well as his Lake
Shore Drive apartments (1951).[67] In New York he visited Skidmore, Owings
and Merrill's Lever Building (1952), the United Nations Headquarters (1952)
and the 'superbly finished' Seagram Building (1958).[68] His visit to Oak Park,
in particular, appears to have shaped his thinking, which is something that he
shares with a number of British contemporaries who had the opportunity to
experience Wright's architecture at first hand.[69]

Following his visit to North America, Wood developed his interested in
timber construction and the potential that it offered for the creation of flexible and adaptable living spaces. He was particularly attracted to the post-
and-beam technique, which allowed the construction of a similar type of
frame to that produced when using steel or reinforced concrete, permitting
doors, windows or wall panels to be positioned in response to the desired
internal layout and avoiding, as he observed, 'the self-contained cubes of the
paralysed plan'.[70]

The houses that Wood went on to design after his participation in the
competition reflect the significance that it had for his practice and this is
most evident in his design for Fenwycks (Chapter 7) and his response to
the steeply falling site on which the Picker House was built (Chapter 10). It
was perhaps as a result of his participation in the competition that he was

Figure 5.11 Kenneth Wood, Forestry Commission District Office Motor Transport Servicing and Workshop, 1968.
Credit: Colin Westwood/RIBA Photographs Collection.

given the opportunity to pursue his interest in timber construction on a larger scale, in one of the few projects that his firm completed outside Surrey, a District Office for the Forestry Commission at Santon Downham in Suffolk, which was given a Civic Trust commendation in 1971 (see Figures 5.10 and 5.11). The Commission required an administrative and service complex to act as the centre for the management of 50,000 acres of forest at Thetford Chase, the second largest forest in England. As a gateway to the forest for visitors, the Commission was keen to demonstrate the use of native timbers in its construction.[71] Wood designed a group of three new buildings around an open grass court, using post-and-beam construction for the administrative building and laminated timber frame construction for the transport unit and fire station. *Design* magazine referred to the group as: 'a well-mannered trio of buildings which fits comfortably into the surrounding landscape and proves, yet again, that decent proportions and sensitive siting are all important when designing buildings in the landscape.'[72]

Notes

1 John Summerson in Trevor Dannatt, *Modern Architecture in Britain* (London: BT Batsford, 1959), 13.

2 The Association became the Timber Research and Development Agency (TRADA) in the 1960s. See, http://www.trada.co.uk/anniversary/index.html, accessed 11 November 2011.
3 Raymond McGrath, *Twentieth-Century Houses* (London: Faber & Faber, 1934), 45.
4 McGrath, *Twentieth-Century Houses*, 120.
5 R. Furneaux Jordan, 'Building in timber: advantages and limitations', *The Times* (3 March 1936): 49.
6 F.R.S. Yorke, *The Modern House*, rev. edn. (London: Architectural Press, 1951), 50.
7 Michael Manser, 'Timber', *Home* (November 1961): 84.
8 David H. Cohen, 'A history of the marketing of British Columbia softwood lumber', *Forestry Chronicle* (September/October 1994): 578–84, see 578–9.
9 Cohen, 'A history of the marketing of British Columbia softwood lumber', 579.
10 David Cohen has suggested that this was due to the fact that there was 'poor follow-up' after the exhibition. See Cohen, 'A history of the marketing of British Columbia softwood lumber', 579–80.
11 Cohen, 'A history of the marketing of British Columbia softwood lumber', 581.
12 Cohen, 'A history of the marketing of British Columbia softwood lumber', 581.
13 Advertisement for Canadian Douglas fir, *Wood* (September 1955): 11.
14 From 1956 *Wood and Wood Products* began to promote the idea of a national forest products week. On the Canadian Wood Council, see: http://www.cwc.ca/AboutCWC, accessed 24 July 2009.
15 The Trend House programme took inspiration from the American Case Study House programme, a landmark in the advance of architectural modernism and the concept of modern living. See, Elizabeth A.T. Smith, *Case Study Houses, 1945–66. The California Impetus* (Cologne: Taschen, 2007).
16 On the trend house programme in Canada see Allan Collier, 'The Trend House Program', *Journal for the Society of the Study of Architecture in Canada* (June 1995): 51–4.
17 Collier, 'The Trend House Program', 51.
18 Collier, 'The Trend House Program', 51.
19 Collier, 'The Trend House Program', 51, 53.
20 As Collier has indicated, the houses also featured new materials such as plastics and plywood. Collier, 'The Trend House Program', 51.
21 Daily Mail Ideal Home Exhibition, Olympia, 1957. Catalogue, 80.
22 Daily Mail Ideal Home Exhibition, Olympia, 1957. Catalogue, 84–5.
23 Daily Mail Ideal Home Exhibition, Olympia, 1957. Catalogue, 84–5.
24 For film of the Canada Trend House see: http://www.britishpathe.com/search/query/ideal+home+exhibition+1957, accessed 28 May 2013. See also, for example, an advertisement for Canadian hardwoods featuring the maple flooring in the lounge of the Canada Trend House. *A&B* (20 May 1958): 2 and *Wood* (March 1958): 4.
25 *AR* (January 1957): iv; *AJ* Supplement (14 March 1957): 11; Astragal, 'Ideal homes again', *AJ* (14 March 1957): 384.
26 Kenneth Wood, in discussion with the author, March 2013.
27 On the Nelson and Parker house, see 'New Buildings, United Kingdom, Timber Houses', *Interbuild* (March 1959): 20–3.
28 British Pathé News Item. Canadian Style Homes, 1958. http://www.britishpathe.com/video/canadian-style-homes, accessed 17 October 2012.
29 British Pathé News Item. Canadian Style Homes, 1958.
30 'Timber house designs', *A&B*, (26 March 1958): 406–11; 'Timber houses off the peg', *AJ* (27 March 1958): 460–5; 'Global report: United Kingdom', *International Prefabrication and New Building Technique*, (April 1958): 240; Margaret Parker, 'Built in Canadian style', *The Lady*, (10 April 1958): 498–9; 'Current topics.

Timber House Exhibition', *Wood* (March 1958): 82; 'A Design in Canadian Timber', *Daily Mail Book of House Plans*, 1962, loose cutting, KWPP.

31 Housing trends, KWPP.

32 There are clear similarities with Wood's design for Wildwood, which slightly pre-dates Design 109, among them the use of mixed materials, the incorporation of a double-height living area and playroom and play terrace.

33 Housing trends, KWPP.

34 Parker, 'Built in Canadian Style', 498.

35 Parker, 'Built in Canadian Style', 498.

36 Parker, 'Built in Canadian Style', 499.

37 'House at Blackheath, London, S.E., designed by Eric Lyons', *AJ* (16 January 1958): 105.

38 Cohen, 'A history of the marketing of British Columbia softwood lumber', 581.

39 See, for example, *Life* (23 March 1959): 97.

40 'Livability unlimited with use of wood', *Quesnel Cariboo Observer* (16 September 1965): 5.

41 'Livability unlimited', 5.

42 'Increased plywood exports maintain industry', *Quesnel Cariboo Observer* (16 September 1965): 4.

43 'Increased plywood exports', 5.

44 'Canadian timber-built houses', *The Builder* (17 July 1964): 139.

45 *Frame Construction with Canadian Timber*, 5. KWPP, undated.

46 *AJ* Supplement (11 August 1955): xxviii; *AJ* Supplement (21 November 1973): 16.

47 Advertisement for Insulight Double Glazing Units, *AJ* Supplement (7 March 1957): 37.

48 Ideal Home Exhibition catalogue (1961): 82–9, 106.

49 These included: 'Timber construction in Canada', *AJ* (2 October 1958); 'Wooden walls', *The Listener* (25 September 1958); 'The role of timber in Canadian build-ing', *The Illustrated Carpenter and Builder* (19 June 1959); 'The structural use of timber', *A&B* (February 1959); 'Canada's architectural development: a view from another world', *Canadian Architect* (February 1959).

50 General notes from Wood's 1958 visit to North America, 2. KWPP.

51 On Vancouver see: R.W. Liscombe *The New Spirit: Modern Architecture in Vancouver, 1938–63*, (Vancouver: Douglas and MacIntyre/Canadian Centre for Architecture 1997); A.C. Elder *et al.*, *A Modern Life, Art and Design in British Columbia, 1945–60* (Vancouver: Arsenal Pulp Press/Vancouver Art Gallery, 2004). Quotation, North American Diary by Kenneth Wood, 2. KWPP.

52 General notes, 2. KWPP.

53 General notes, 2. KWPP.

54 General notes, 3. KWPP.

55 General notes, 5. KWPP.

56 'Romance and commonsense. The story behind Canadian timber homes', 2. Notes for an article for *New Homes Magazine*. KWPP.

57 'The role of timber in Canadian building', 2. Notes for an article for *The Builder*. KWPP.

58 General notes, 4. KWPP.

59 General notes, 3. KWPP.

60 General notes, 3, 5. KWPP.

61 Collier, 'The Trend House Program', 51.

62 Sherry McKay, 'Western living, western homes', *Society for the Study of Architecture in Canada Bulletin* 14/3 (September 1989): 65–73, on 73.

63 'Canada. Architect's own house in West Vancouver. John Porter', AD (March 1956): 91–2.
64 Spence cited in R.W. Liscombe, *The New Spirit*, 117.
65 General notes, 5. KWPP.
66 Kenneth B. Wood, 'Timber construction in Canada', *AJ* (2 October 1958): 485–92.
67 General notes, 6. KWPP.
68 General notes, 6. KWPP.
69 As James Stirling observed if 'you can't see a FLW building you must be in it.' See Mark Girouard, *Big Jim, The Life and Work of James Stirling* (London: Chatto & Windus 1998), 45.
70 'Wooden walls. Reflections by Kenneth Wood on the architecture of British Columbia', *The Listener* (25 September 1958): 459.
71 The site, a large clearing within the forest, contained a number of buildings in the local style of brick and knapped flint, which formed a significant part of Santon Downham village, along with several old Nissen huts.
72 'Offices among the trees', *Design* 287 (November 1972): 69.

6   Exhibited house
Vincent House, 1959

Vincent House was designed for a sloping plot of just under an acre that had previously formed part of the grounds of a large mansion at Kingston Hill in Surrey. The house was created as a replacement for the Manor House at Stoke D'Abernon that had been home to Wood's clients, the Vincent family, over many years, but had become uneconomical to maintain (see Figure 6.1). Two years after its completion, the house was exhibited at the Architecture Today Exhibition in London. Sponsored by the Arts Council of Great Britain and the Royal Institute of British Architects, the exhibition took place at the Arts Council Galleries in St James Square from 28 June to 29 July 1961 and was timed to coincide with the Congress of the International Union of Architects that was held in London in that year. In the same year the house was exhibited within the quite different context of Bentalls department store at Kingston, as part of Architecture Week, an event organised by the Kingston upon Thames District chapter of the South Eastern Society of Architects. The two exhibitions situated Wood's work for different audiences. While Architecture Today aimed to promote the achievements of British architects to international visitors, Architecture Week had a strong local focus and was developed with the aim of showcasing work in order to attract commissions.

Rather than exhibiting a representative selection of projects, the organising committee for Architecture Today took the decision to show what was, in their opinion, Britain's best contemporary architectural design and invited a number of leading firms to participate in addition to issuing an open call for projects.[1] Wood was not among the elite and made his entry via the open submission process, putting forward three buildings for consideration: Emmanuel Church at Tolworth, Wildwood at Oxshott and Vincent House at Kingston, from which Vincent House was selected.[2] Following its London opening, Architecture Today toured through 1961 and 1962, travelling first to Thomson House at Cardiff and then to local art galleries and museums (Bedford, Norwich, Swansea, Southampton, Huddersfield, Leicester, Eastbourne, Hull, Kettering, Newcastle and Sheffield) and to educational establishments (Edinburgh College of Art, Nottingham University, The School of Architecture in

Figure 6.1 Kenneth Wood. Vincent House, Kingston Hill. Garden elevation, 1959.
Credit: RIBA Library Photographs Collection.

Leeds, Mid-Herts College of Further Education and The Campus in
Welwyn Garden City).[3]

Around fifty projects were chosen for the exhibition, which featured work
by Ernö Goldfinger, Powell and Moya, Yorke, Rosenberg and Mardall, and
Fry, Drew and Partners, among others. As J.M. Richards noted in the exhibition catalogue, the selected work reflected the breadth of architectural practice in Britain at that time.[4] The residential category, in which Vincent House
was exhibited, included local authority housing and speculative schemes
and featured low-rise, high-density suburban and village housing alongside
Corbusian-inspired urban mass housing: James Stirling and James Gowan's
flats at Langham House Close, Ham Common; the London County Council's
Alton Estate (West) in Roehampton; Eric Lyons' houses at Corner Green
in Blackheath and Highsett flats at Cambridge; the Park Hill development
in Sheffield; a terrace of six houses at South Hill Park in Hampstead by
S.F. Amis, W.G. Howell and G. Howell; village housing and a clubhouse
at Rushbrooke in Suffolk by R. Llewelyn Davies, John Weeks and Michael
Huckstepp; luxury flats at St James's Place and a 'cluster block' at Bethnal
Green by Denys Lasdun and Partners; Regent's Park housing redevelopment

at St Pancras by Edward Armstrong and Frederick MacManus; and housing at Peterlee New Town by Peter Daniel and F.G. Nixon with Victor Pasmore. Several examples of student accommodation were also included alongside designs for new schools and colleges.

As the list of participants indicates, many of those whose work was included were well-established figures, making the selection of Vincent House a breakthrough for Wood's firm and an acknowledgement of the quality of the work that it was producing.

The exhibition catalogue had a material and aesthetic emphasis and concrete was much in evidence in the twenty-four projects depicted. It was not, according to Kenneth J. Robinson, writing in the *Spectator*, an exhibition for the layman. Although he appreciated J.M. Richards's catalogue essay, Robinson commented that Architecture Today was

> so undercaptioned and so badly catalogued that it can be used only as a starting point for study. You would have to be very interested indeed not to be put off by the catalogue's description of a block of flats in Bethnal Green where 'the vertically segmented form reduces the apparent mass and repetitive content'.[5]

Vincent House was one of several private houses included in the exhibition, none of which made it into the photographic section of the catalogue. They included a weekend house of timber construction at Fawley Bottom, near Henley on Thames by Lionel Brett; a brick and timber construction, single-storey courtyard house at Hampstead by Trevor Dannatt; a two-storey brick and steel-framed house at Hampstead by H.C. Higgins and R.P.H. Ney and Partners; at Bromley, a timber-framed house with linked pavilions, by W.G. Howell and J. Killick; a two-storey brick house at Highgate by Leonard Manasseh; High Sunderland, the single-storey timber-framed house near Galashiels that Peter Womersley had created for textile designer Bernat Klein; and a house in Kano, Nigeria by Architects' Co-Partnership.

Vincent House was constructed using a precast reinforced concrete frame, which was an unusual choice of materials for Wood at the time (see Figure 6.2).[6] Writing some time after the exhibition, Ian Nairn described it as:

> Bigger than most English post-war houses, and with the right kind of opulence – not assertive, but rich and self-confident. Split pitched roof which gives clerestory lighting to the first floor; big two-storey balcony in front. Comparison with an up-to-date house of the 1930s shows how much more supple and fluent modern design has become; it is no longer obliged to prove anything, it can just be itself.[7]

Wood developed the design in response to mature trees on the plot and a change in level across its width, using a modular plan to allow the incorporation of some prefabricated elements. At ground floor level the concrete

Figure 6.2 Kenneth Wood. Vincent House, Kingston Hill. Exterior detail showing concrete frame and entrance to the self-contained first floor accommodation, 1959.
Photographer: Douglas Whittaker. Credit: RIBA Library Photographs Collection.

frame was filled with panels of glazing and handmade brick and at first floor level with lightweight concrete blocks.[8] Windows were either in aluminium (sliding sashes) or mahogany frames, with black slate cills. The first floor of the house was faced with horizontal cedar boarding and a balcony on the garden side extended its full width, giving shade to the living space below.[9]

The house was planned to accommodate an extended family: Lady Vincent, her daughter Pamela, who had been a dancer with the Ballet Rambert, her son-in-law Douglas Whittaker, principal flautist with the BBC Symphony

Figure 6.3 Kenneth Wood. Vincent House, Kingston Hill. Master bedroom with
       view to en suite bathroom, 1959.
Credit: RIBA Library Photographs Collection.

Orchestra, and their young children.[10] The house was arranged on an
L-shaped plan, with a sheltered entrance court and driveway and provided
3,400 square feet of living accommodation.[11] The smaller wing housed a
self-contained first floor apartment for Lady Vincent (living room, bathroom,
kitchenette, bedroom and balcony) below which was a carport with space for
two cars. The flat, which had independent access via an external staircase,
was connected to the main wing of the house by way of a covered balcony
(see Figure 6.2). The first floor of the main wing contained four bedrooms,
a bathroom, and a master bedroom with an adjoining dressing room and
shower (see Figure 6.3). Removable walls between three of the bedrooms
anticipated their future arrangement as two larger rooms. With the exception of the bathroom and the smallest guest bedroom, all of the first floor
rooms were south-west facing and overlooked the garden. The landing was
spacious and was designed to be used as an occasional art gallery. Finished
in rough-textured plaster, the plain walls were intended to provide a simple
background for the display of pictures.[12] The area was well-lit, with natural light from a large window, for which Wood designed a window seat,
and from clerestory lighting above (see Figure 6.4). Timbers were exposed to

Figure 6.4 Kenneth Wood. Vincent House, Kingston Hill. First floor landing with window seat, 1959.

Photographer: Douglas Whittaker. Credit: RIBA Library Photographs Collection.

reveal the scissor-beam construction of the roof and the roof voids above the guest bedroom and bathroom were used as secondary storage areas.

On the ground floor, to the courtyard side, were a kitchen/breakfast room, a playroom, and a workshop and dark room that Douglas Whittaker used for photography – a hobby that he pursued to a professional standard. To the rear, a south-west facing living area opened onto the garden. Writing in *Country Life*, Mark Girouard felt that the house had been 'built for people of a practical turn of mind' and observed in its interiors 'numerous ingenious gadgets and fittings, the result of fruitful co-operation between client and architect'.[13]

As a professional musician, Douglas Whittaker required a suitable environment in which to rehearse, and in their brief to Wood the family also indicated their desire to hold musical performances at home. Writing in 1934, Raymond McGrath, who had first hand knowledge of developments in acoustics from his work at the BBC, noted the general lack of attention that architects were giving to that aspect of domestic design: 'The living-room, which is generally the music-room in addition, is frequently not designed with a view to its sound qualities, though the science of sound now gives us full knowledge of the desired effects.'[14] By the 1950s that had begun to change. An article on Peter Womersley's Farnley Hey, which appeared in *The Architectural Review* in 1955, reflects the growing emphasis on sound that resulted from the advance of open-plan living, the experience of those spaces, and the growing interest in audio equipment and recorded sound quality that had begun to emerge among affluent male consumers. The journal noted:

> Accommodation had to be made for both live and recorded music. A music gallery was designed for the playing of the former, and the apparatus for recorded music is housed between the columns separating living room and study. The quality of sound produced is helped by the height of the room with its ceiling panelled in mahogany, by the area of stone paving immediately below the amplifier unit, and by possible alterations in the texture of the wall opposite to the speaker – plate glass or curtain in varying degrees.[15]

Wood's Picker House, of the mid 1960s (Chapter 10), is similar in design, with a music gallery above the main living area and a specially designed, built-in audio fitment, located between timber elements on the floor below.

In his design for Vincent House, Wood made a feature of the fall in land across the plot, using it to create a dais to one end of the living area to accommodate a formal dining area and space for the Whittaker's piano and audio equipment. Timber was used throughout the interior for its natural acoustic qualities and its appearance, durability and ease of maintenance. The music and dining area had a boarded wall in red cedar and a floor in African muhuhu. A raked timber ceiling was introduced to enhance its acoustic performance (see Figure 6.5). Photographs of the music dais, taken not long after the house was completed, depict a large carpet, or wall-hanging to the rear of the piano, which may also have been introduced for acoustic purposes. Sapele mahogany was used for window frames, doors and built-in furniture. The ceilings in the living area were in British Columbian pine and the staircase in mahogany with a bronze and plate-glass balustrade.

Two years before the Architecture Today exhibition, Vincent House featured alongside Farnley Hey in *Home* magazine, in an article written by Michael Manser on the new trend for timber; and in the year of the exhibition the house was included in an article on the care and use of wood in *Good Housekeeping* magazine.[16] The article in *Good Housekeeping* reflects

Figure 6.5 Kenneth Wood. Vincent House, Kingston Hill. Music and dining area, 1959. Photographer: Douglas Whittaker. Credit: RIBA Library Photographs Collection.

the values associated with the use of different woods within British domestic interiors at that time, extending from the fashionable exoticism of African hardwoods to the cooler modernity that was associated with Canadian and Scandinavian timbers such as maple and pine.[17] Wood's choices for Vincent House complemented the antique furniture that was already owned by his clients. Writing in *Country Life*, Mark Girouard described the effect in the main living area:

> The furnishings of the room are a pleasant mixture of the old and the new, the useful and the decorative: two portraits by Romney and some modern pictures; furniture both modern and old, and comfortable chintzy chairs; a grand piano on the platform and a built-in gramophone and tape-recorder unit.[18]

Wood paid careful attention to the design of the outdoor spaces of the house and its thresholds and boundaries. To the front of the house an entrance court served the needs of the car owner and incorporated spaces for planting, to create a welcoming environment for visitors. A gently sloping path, with a raised flower bed beneath, passed in front of the kitchen

Figure 6.6 Kenneth Wood. Vincent House, Kingston Hill. View to kitchen and breakfast area, 1959.
Photographer: Douglas Whittaker. Credit: RIBA Library Photographs Collection.

and breakfast room windows, connecting the canopied front door with the carport and stairs to the first floor apartment (see Figure 6.6). A pergola enlivened the route from the garden to an outdoor store and connected the house to its boundary wall. To the left of the front door, a high curving wall, designed to preserve a mature tree on the plot, formed the boundary with the entrance court. Set back slightly from the facade to create an additional planting space, it ensured the privacy of the rose garden to its rear. Tall grasses and climbers were originally planted to the driveway side, while planting on the garden side was lower. The rose garden, which was intended as the future site of swimming pool, was set out quite simply, with precast concrete flags and rose beds in between.

The inclusion of Vincent House in the Architecture Today exhibition helped to position the work of Wood's practice at a national level and within a professional context. Although the local architectural exhibitions in which the firm participated often merited no more than a few lines in the main British architectural journals, they represent a significant means through which its work was disseminated to potential clients. Records of some of the architectural exhibitions held at Kingston upon Thames in the 1960s give an indication of the architectural landscape in Surrey at that time, the type of work that local practices were engaged in, and the ways in which the relationship between modern architecture and the modern home was being shaped within a local retail context.

Following the success of an Architecture Week held at Guildford in 1960, the Kingston upon Thames district chapter of the South Eastern Society of Architects arranged its own exhibition of photographs, drawings and architectural models at Bentalls, the Kingston department store, from 9–14 January 1961. The market town of Kingston had a thriving retail centre and drew its custom from a large and affluent surrounding population. Since the 1920s, Bentalls had sought to attract motorists to the store from rural parts of Surrey, incorporating extensive covered car parking and restaurant facilities for those making longer journeys.[19] The store was a significant presence in the town and from the 1930s developed an emphasis on housing and household goods.[20] Capitalising on the rapid development of surrounding areas, Bentalls also moved into property sales, launching an estate agency that initially marketed houses on the new estates that were being built at Chessington, for which it furnished show homes to promote the services of its furnishing department.[21] In the 1950s and 1960s the store ran a variety of exhibitions and events including Italy in Kingston (1955), an American Fair (1960), and a Britain's 'Best at Bentalls' event (1968) in conjunction with the Council of Industrial Design.[22]

Architecture Week was held at the store's Wolsey Hall and was organised by Kenneth Wood, Alan Blanc, R.A. Michelmore and James Lomas.[23] The foreword to the exhibition catalogue emphasised the social role of the architect: 'No-one can ignore the effects of building; architecture is a social art which in time creates the whole environment in which we live, learn, work and play.'[24] The exhibition was opened by Sir Hugh Casson, former Director of Architecture at the Festival of Britain who was at that time Head of the School of Interior Design at the Royal College of Art (see Figure 6.7). The store's willingness to support an architectural event was perhaps not unrelated to the personal interests of its Managing Director, Gerald Bentall. Witley Park near Godalming in Surrey, the modernist house designed for him by Patrick Gwynne (and Gwynne's largest private commission) was completed in the following year.[25]

Twenty-seven individuals and groups participated in Architecture Week and with the exception of seven London-based firms all were located within a few miles of Kingston, an indication of the vibrant local architectural scene at that time: Robert Bailie (East Molesey), Barber, Bundy and Greenfield (Kingston), Alan and Sylvia Blanc (Coombe), the Architects' Department at Messrs Currys (Cobham), Jack Godfrey Gilbert (Wimbledon), Fred Greenwood (Richmond), J.E.K. Harrison (Wimbledon), E.R. Heathcote (Cobham), W.G. Jones of Powell and Alport (Croydon), the Architects' Section of the Borough Surveyor's Department at the Royal Borough of Kingston upon Thames (Kingston), James Lomas (Weybridge),

Figure 6.7 Kenneth Wood, Rowan Bentall and Sir Hugh Casson at Architecture
          Week at Bentalls department store, Kingston, 1961.
Credit: Courtesy of Kenneth Wood.

Eric Lyons (East Molesey), Michael Manser (Leatherhead), Noel Moffett
(Kingston), John Strubbe (Richmond), A.L. Tamkin of the Milk Marketing
Board (Thames Ditton), Thomson and Gardner (Sutton), W.W. J. Trollope
(Tolworth) and Kenneth Wood (East Molesey).

  Fifty-two projects were exhibited, thirty-two of which were for individual houses or housing schemes. Residential projects in south-west London
and Surrey included: Alan and Sylvia Blanc (house at Coombe Hill), the
Architects' Department at Messrs Currys (house at Winnall); John Dodridge
(house at Thames Ditton); Fred Greenwood (British Airways Staff Housing
Society at Stanmore), the architects' section of the Borough Surveyor's
Department at Royal Borough of Kingston upon Thames (old peoples'

dwellings, Cambridge Gardens, flats at Chessington, flats at Elm Road, maisonettes at Richmond Park Road); James Lomas (house at Weybridge); Eric Lyons (Parkleys at Ham Common, Fieldend at Teddington); Michael Manser (house at Epsom, bungalow at Leatherhead); R. Michelmore of Architects Co-Partnership (housing estate at Wandsworth); Noel Moffett (house at Esher); Thomson and Gardner (flats at Sutton); and G.B.A. Williams (Calkin House at Cheam).

Wood exhibited several of his own projects – Design 109, his BCLMA competition house, Vincent House, Wildwood, two parish halls at St Mary-at-Finchley, Oxshott Village Centre and Emmanuel Church. His friend Peter Jones created a mural for the event, which the *RIBA Journal* noted as a 'substantial' exhibition and 'a useful form of advertising the profession'.[26] Within the exhibition hall, projects were presented as plans, drawings and models and were interspersed with small room sets that incorporated items of furniture that could be purchased within the store.

A larger architectural exhibition, Design 66, was held at Bentalls' Wolsey Hall in 1966. The exhibition organiser, Hugh Smart, was assisted by Alan Blanc, Ray Palmer, Jim Lomas and Ivor Plummer. On that occasion, sculpture and ceramics by Janusz Lewald-Jezierski were displayed alongside the architectural works. The exhibition was of significant scale, including around 300 projects by members of the South Eastern Society of Architects. Among the exhibitors were Scott Brownrigg and Turner, Michael Manser, Alan and Sylvia Blanc and Eric Lyons. Wood exhibited three private houses, Oriel House, Torrent House and Dykes, and a chapel at St Paul's Church at Kingston.

The relationship between architecture and retail continued in 1969, when the South Eastern Society of Architects exhibited at the Habitat store in the town's new pedestrianised shopping centre, Eden Walk.[27] Terence Conran established Habitat in 1964 and the branches at Kingston and Manchester were the first to open outside London. The change of venue to a design-led retailer with a young and cosmopolitan vision for the home perhaps suggests that those involved in its planning were alert to its potential to attract an audience in sympathy with modern architectural design. The exhibition incorporated 108 projects by fifty individuals and firms. Among the exhibitors were Norman and Dawbarn; Broadway and Malyan; Dry, Halasz, Dixon Partnership; Noel Moffett and Associates; B. and N. Westwood, Piet and Partners; and Derek Lovejoy and Associates.[28] Wood exhibited four projects that reflect the new commercial areas into which his practice had begun to move: Torrent House at Hampton, a child guidance unit at Wandsworth, showrooms at 64 Borough High Street, and new headquarters for the Forestry Commission at Santon Downham in Suffolk.[29]

Local exhibitions provided valuable opportunities for professional collaboration and exchange between architects as well as occasions for them to evaluate their work alongside that of other practices operating within the same professional arena. They were undoubtedly helpful in raising the profile of Surrey architects and of contemporary architecture more generally.

The extent to which that translated into new commissions is unclear. While each of the local exhibitions in which Wood participated included a wide range of public, commercial and residential projects, the retail context in which the work was displayed was one in which domestic values were prioritised in contrast to the formal and technical concerns of Architecture Today.

Notes

1   Architecture Today exhibition correspondence and minutes. Arts Council of Great Britain Records, 1928–97, Hayward Gallery Material, 1935–97, Exhibition Files, 1945–95, Architecture Today Exhibition 1959–63 (Victoria and Albert Museum, London, Archive of Art and Design, ACGB/121/28).
2   Architecture Today exhibition correspondence and minutes.
3   See, *RIBA Journal*, 63/8 (June 1961): 296. See also, Architecture Today exhibition correspondence and minutes.
4   J.M. Richards, 'Extracts from the exhibition catalogue by J.M. Richards, CBE', *RIBA Journal* (June 1961): 296.
5   Kenneth J. Robinson, 'Fairs and squares', *The Spectator* (21 July 1961): 17.
6   The precast structural frame was supplied by Wadcrete Limited. The house featured in *Concrete Quarterly*. 'Concrete for the house of the future', *Concrete Quarterly* 52 (January to March 1962): 22.
7   Ian Nairn, *Modern Buildings in Britain* (London: London Transport, 1964), 105.
8   'Concrete for the house of the future', 22.
9   As Geraint Franklin has observed, there are echoes in this garden elevation of Eric Lyons' Parkleys Parade at Ham Common. Geraint Franklin, Vincent House, Warren Road, Kingston: A Report by the Architectural Investigation London Team, December 2005. English Heritage Reports and Papers Series B/014/2005, 5.
10   The Manor House, a Grade II* listed Palladian villa, dating from the mid eighteenth century, was built around an older Tudor house.
11   In 2006 an attempt to list the house failed and permission to demolish it and build two detached five-bedroom homes on the site were approved. Royal Borough of Kingston upon Thames, Maldens and Coombe Planning Sub-Committee, 16 November 2006, minute: 28/1.
12   The family intended to hold small, informal exhibitions of art at home and this formed part of the original brief to Wood.
13   Mark Girouard, 'A house built for a musician', *Country Life* (7 June 1962): 1385.
14   McGrath's work on studios for BBC Broadcasting House in London (from 1931) made him alert to developments in this area. The application of professional knowledge to the private space of the home was considered more widely as open-plan living became more popular in Britain and in America. George Nelson and Henry Wright, for example, included a chapter on 'Sound Conditioning' in their book *Tomorrow's House* (1945) in which they discussed the acoustical design of professional broadcasting studios and the application of knowledge to the design of domestic environments. See, George Nelson and Henry Wright, *Tomorrow's House: A Complete Guide for the Home-Builder* (London: Architectural Press, 1945), 143–50.
15   'House in Yorkshire. Architect: Peter Womersley', *The Architectural Review* (December 1955): 364–5.

16 Michael Manser, 'Timber: new trend', *Home* (October 1959): 78–83. 'The wonder of wood', *Good Housekeeping* (February 1961): 56–61, on 56.
17 'The wonder of wood', 56–61, on 56.
18 Girouard, 'A House Built for a Musician', 1383.
19 Rowan Bentall, *My Store of Memories* (London: W.H. Allen, 1974), 82.
20 The firm's Building Department was engaged to build 144 council houses for Kingston Corporation at Norbiton and was also awarded a large contract to complete work at Buxted Park, the home of Mrs Basil Ionides. In the 1930s it also manufactured timber-framed bungalows. Bentall, *My Store of Memories*, 162–4.
21 Bentall, *My Store of Memories*, 163.
22 Bentall, *My Store of Memories*, 264–71.
23 The Wolsey Hall was introduced in the 1930s as the Mannequin Hall and was initially used for functions and mannequin parades. Bentall, *My Store of Memories*, 134–5.
24 Architecture Week catalogue, January 1961. KWPP.
25 Gerald Bentall was Managing Director of Bentalls from 1942–63.
26 'Architecture Week at Bentall's, Kingston', *RIBA Journal* (February 1961): 143.
27 The exhibition ran from 15–22 March. Architecture at Habitat exhibition catalogue, March 1969. KWPP.
28 The full list of participating firms: Eric Lyons and Partners; Garner, Preston and Strebel; Jackson and Edmonds; Sidney Kay, Eric Firmin and Partners; Norman and Dawbarn; Broadway and Malyan; Robin Moors, Allnutt and Partners; Gordon Russell; G.B.A. Williams Partnership; Dry, Halasz, Dixon Partnership; Kenneth Wood; James Parsons; Michael G.D. Dixey (and Michael G.D. Dixey and Richard Suddell); Hutchison, Locke and Monk; Broadbent, Hastings, Reid and Todd; J.E.K. Harrison and Associates; William Crabtree and Jarosz in collaboration with G.C. Fardell, D. Galloway, W.G. Plant; Noel Moffett and Associates; T.B. Bush and Partners; Maxwell New Associates; B. and N. Westwood, Piet and Partners; C. Wycliffe Noble; J.H. Lomas (Borough Architect); Manning and Clamp; Phippen, Randall and Parkes; Hughes and Polkinghorn; Alan and Sylvia Blanc; Gerard Brigden; Pinckheard and Partners; Fitzroy, Robinson and Partners; Burns, Guthrie and Partners; M.A. Munro; Kenneth Adams and Partners; Robert Paine and Partners; Building Design Partnership; Bader and Miller; A.J. and L.R. Stedman; Hubbard Ford and Basil Spence; Derek Lovejoy and Associates; Charles Pike and Partners; John Spence and Partners; John Voelcker; Scott, Brownrigg and Turner; Scherrer and Hicks; Darbourne and Dark; Michael Manser Associates; W.S. Hatrell and Partners; Bertha Roake; Raymond Ashe (Surrey County Architect); Robert Bailie.
29 Architecture at Habitat exhibition catalogue, March 1969. KWPP.

Fenwycks, a split-level, half-timber house at Camberley in Surrey, is a British example of a fully realised developing home: a house designed for planned alteration to meet the anticipated long-term needs of its occupants (see Figure 7.1). It is one of a number of examples of developing houses designed by British architects in the 1950s and 1960s and offers an opportunity to consider the way in which the concept, which was well established in North America, evolved Britain in that period.[1] The house exemplified the values of flexibility, adaptability and comfort that were central to the Canadian lumber industries' marketing efforts in Britain, and was used by the Council of Forest Industries of British Columbia to promote timber construction.

Writing during the war, the architect and town planner C.H. Reilly considered the attraction of modern domestic architecture and its future:

> The young man or woman … with no particular social ties to a neighbourhood, may well demand the lighter, gayer type of structure, 'perching on the ground like a bird rather than growing out of it' – as Corbusier's houses have been described. For such a free, 'modern' design constructed in light steel, wood or ferro-concrete, with the overhanging parts these alone can give, answers to the semi-outdoor life such people desire. This kind of house, too, is often little more than a temporary abode, a hardly more permanent possession than the motor car or aeroplane with which the latter has so distinct an affinity. If it is to be a place for rearing children, it may well be a very healthy one, but it is hardly a symbol of an established family, nor probably desired as such.[2]

In his design for Fenwycks, Wood responded to his clients' desire to create a permanent setting for family life, using timber to create a light, adaptable building that challenged its association with temporary structures and with the values of freedom and mobility with which Reilly associated it. The house can be situated within a longer history of design for adaptation.

During the depression era, American architects and house-builders offered a variety of pay-as-you-build housing solutions to financially constrained families. Sears, the mail-order supplier, marketed houses for staged

Figure 7.1 Kenneth Wood. Drawing of Fenwycks, Camberley, *c.* 1959.
Credit: Courtesy of Kenneth Wood.

development in the 1930s. The Homecrest, one of its Honor-Bilt house
designs, was a five-roomed house planned for extension with a porch, a
garage wing and a third bedroom.[3] Among many other examples of houses
for staged construction that featured in American publications and exhibitions of the 1930s are a small house by New York architect Leigh French
that appeared in *House and Garden* magazine in 1931 as an example of
a viable way of building to stimulate the economy, and a 1937 exhibition
house by Vincent Schoeneman, for Pittsburgh's Better Homes Show, which
aimed to address the problem of affordable house-building for the 'average
income family'.[4] In the early 1940s, while working for Frank Lloyd Wright
at Taliesin, Loch Crane developed the idea of an 'expandable house' and
went on to test it in a house built for his family at Point Lorna, San Diego
after the war.[5] A widely disseminated example from the late 1940s is Marcel
Breuer's exhibition house, built in the garden of the Museum of Modern
Art in New York in 1949.[6] The house was, like Fenwycks, designed as a
bespoke family dwelling rather than a prototype for volume construction
and Breuer's imagined client – 'the commuter who has personal views in
selecting his land, probably at least an acre' – was much like Wood's.

Breuer's expandable house was designed to be built in two phases for a family with young children. In the initial phase the living space was arranged on one floor, as a living-dining room, two bedrooms, a play room with its own entrance, and a bathroom, kitchen, and utility room. The phase two house, planned for completion when the children were older and the family more financially able, incorporated an additional garage-storage area with a bedroom, bathroom and sun terrace above. Breuer intended that:

> In the second phase, if there are more than two children, they can take over the master bedroom of the first phase and use the children's play-room as their own living room or study. They are near the living room and kitchen, easily supervised, and yet they are separated.[7]

The kitchen was envisaged as the hub of the house and was described as follows:

> The kitchen is central, controlling all activities. There is an observation glass panel towards the playroom. Kitchen shelving and cabinet work is simple, partly open (since there is usually no dust in the country) with sliding doors of the simplest construction instead of swinging doors. Counters are of hardwood. Kitchen and utility room and service yard are adjacent and equipped so that housework is reduced to a minimum. This was a very definite principle in planning the whole house and in selecting the materials and finishes, especially for the floors, walls and for all the furnishings. The utility room can serve also as an emergency bedroom for night sitters, for occasional help, etc. A radiant heating system in the floor slab is planned to make the summer-cool stone floor warm in the winter. The extensively used bluestone flagging is a very practical floor to maintain, especially in rooms on the garden level.[8]

The outdoor areas of the house were separated into distinct zones by means of 'freestanding louver partitions or bench-like stone walls' and included: a parking area to the front of the garage; a patio entrance terrace; a service yard; garden areas adjoining the bedrooms; a play area for the children; and adjoining the living-dining room a terrace with an outdoor grille.[9]

Breuer's expandable design focused on the family in its growth phase. A slightly later example, the Flexabilt House by Frank Robertson – several hundred of which were built at San Antonio in Texas in the 1950s – illustrates a housing developer's approach to the design of houses for long-term occupation.[10] In response to his observations of the housing market and his personal experience as a home owner, Robertson recognised the value of planning for reduction as well as expansion in family size.[11] The Flexabilt House – a single-storey house of 1,250 square feet – was arranged to function as two self-contained apartments – a kitchenette apartment and a two bedroomed apartment – or as a large single-family unit configured for different patterns of occupation over an extended period.[12] In 1953, the house featured in an article in *Popular Mechanics* magazine, as a 'lifetime home' in which a young couple might live first in the kitchenette apartment, offsetting their mortgage by renting out the larger space, reversing the arrangement with increased financial ability and family growth, before combining the apartments into one large family unit. Finally, once the children had left home, it was envisaged that the apartment would revert to its original form as two smaller dwellings.[13]

Britain's immediate post-war housing and labour shortages created a similar context in which staged development became attractive to architects and house buyers. In 1946, working within the parameters of restrictions on the size of house that could be built, Ernö Goldfinger designed a pair of semi-detached houses with a removable party wall to allow their future occupation as a single dwelling.[14] A number of other examples of expandable houses appeared in British architectural journals and publications of the early 1950s and reveal the ways in which regulatory constraints had begun to stimulate architects to design houses for expansion in anticipation of improving economic conditions.[15] Among those included in F.R.S. Yorke and Penelope Whiting's *The New Small House* (1953) were a house at Kingston in Surrey by Tayler and Green, which had wood framed end walls that could be dismantled and re-used within an enlarged building; and one by Stephen Gardiner, at Fitzroy Park in London, which was planned to accommodate a penthouse addition, for which provision was made in the roof slab to incorporate a stair opening to a new upper floor.[16]

Slightly earlier than Wood's Fenwycks, Stirling and Gowan's Quadrant House of 1957, which was commissioned by *House and Garden* magazine as a house 'for young marrieds whose ambition outruns their bank balance' responded to the question of financial resources as well as that of changing family needs.[17] Their design featured in the magazine's April 1957 edition and the accompanying editorial commented on the challenge that its staff had posed the architects:

> The problem, then, is to build a house which can be added to in stages which will appear an architectural entity at each step, and which is capable all its life of 100% efficiency, with no overcrowded or empty rooms. This is an idea for such a house.[18]

The first stage, for a single person or a couple without children, had an unconventional layout with a central two-storey core containing services (w.c., boiler, linen-cupboard cooker, sink, shower and store) and a multi-purpose ground floor room with an upper level room for sleeping, study or storage.[19] The addition of three further quadrants, to create a bedroom and garage, a larger kitchen and dining room, and an expanded living room and further bedroom, allowed for its development in line with family growth. Stirling and Gowan's design also permitted the rearrangement of the space as a smaller unit, allowing an older couple to sublet part of their home to provide additional income in retirement. All of the structural walls were planned for completion at stage one, with the infilling and roofing of the quadrants taking place in three successive stages, partition walls allowing the reconfiguration of the interior with the addition of each quadrant. The total cost was estimated at £4,850 in affordable stages of £1,800, £800, £1,300, £950. Quadrant House is one of few examples of the 1950s that looked beyond family growth to consider family lifecycle.

The publication of *Homes for Today and Tomorrow* (1961), which outlined the findings of the Parker Morris Committee, recognised the need to take a longer-term view of public housing design and to consider the ways in which the needs of occupants might change over time.[20] The report responded to recent experiments in the design of long-term living environments and pointed to the need for further research:

> It is not to be wondered at that the adaptable house – the house which could easily be altered as circumstances changed – is a recurring theme in the evidence we received and in our own thoughts. At the present stage of development, such a dwelling is some way from practical reality, because of the high cost and other difficulties … We see the investigation of the practical possibilities of doing it easily and at reasonable cost as one of the most important lines of future research into the development of design and structure. The sooner it is started the better.[21]

In the same year 'The Growing Home', an exhibit designed by Gordon and Ursula Bowyer for the Design Research Unit, was displayed at the BFM Furniture Exhibition to demonstrate how a house could 'be altered to adapt itself to the needs over many years of a growing family'.[22] In 1962 architects at the British Ministry of Housing and Local Government devised a house with an adaptable plan for use over a period of fifty years and with reference to an anticipated seven-stage family lifecycle.[23] In H. Dalton Clifford's *Houses for Today*, which was published the following year, eight out of the thirty-eight examples were designed as family houses with extension in mind.

Fenwycks gave Wood the opportunity to put his own thinking in this area into practice with clients, Mr and Mrs Eric Paton and their three small children, who shared his belief that a carefully conceived house could provide a comfortable home for life. In terms of their immediate requirements, the Patons wanted a comfortable family home with plenty of room for home entertaining. They wanted to be able to increase the size of their home in the future and were interested in the possibility of creating additional self-contained accommodation within it, for the use of their children, should any of them wish to stay at home as young adults, or older family members, should the need arise.

Designed in 1959, Fenwycks was built by a local contractor, Dalley & Sons of Camberley, and was completed in the following year at a cost of around £6,500 excluding land.²⁴ The project was published in architectural and technical journals and consumer magazines, among them *Architectural Design, The Illustrated Carpenter and Builder, Modern Woman, House Beautiful* and *Votre Maison*.²⁵ Writing in *Home* magazine in 1961, Michael Manser described Fenwycks as 'a house that stands squarely half-way between expansion and contraction'.²⁶ At the time of its completion in 1960, Fenwycks had a floor area of 2,000 square feet and was designed to allow the rearrangement of certain interior elements as well as simple modular extension of the building. Wood's design made a feature of the sloping site, which had open views to the south and south-east and was protected by trees to the north. A porch, entrance lobby, hall, garage, store, living area and adjoining patio were situated at ground level. A kitchen and dining area were located at mezzanine level, with four bedrooms and a bathroom half a floor above. The lower ground floor wing of the house was built in brick and the first floor was of timber platform frame construction, with the framing in Douglas fir and Pacific hemlock and exterior weatherboarding and panelling in fir. For the upper living/dining area, which was designed to simplify extension of the building in the future, Wood employed post-and-beam construction. The exterior spaces of the house, including the layout of the garden, were planned to require minimal alteration at the time of the building's extension.

Wood designed the split-level living area to allow its use as a single space for large social gatherings while maintaining a sense of scale conducive to family life. The entrance hall was planned as an open zone within the lower living area, where it could function as a reception space or as part of the main living area at times when additional room was needed. A separate glazed draught lobby, containing a cupboard for deliveries and one for outdoor clothes, served as a compact but adequate entrance when the main hall was given over to entertaining. The living area contained various screening devices, allowing its separation into smaller, more intimate zones when the family was at home alone. A venetian blind behind the central chimney allowed the visual separation of an enclosed seating area around the fireplace, which was of strikingly modern design, with a bright yellow flue, enclosed in a tube of spun concrete, surrounded by white painted brick and a tiled hearth (see Figure 7.2). *House Beautiful* described the Patons as 'a cheerful, gregarious family' and saw in Fenwyck's design a reflection of their hospitable and open nature: 'it's a friendly, welcoming house, unrestricted by unnecessary walls and doors'.²⁷

The kitchen, which was fairly narrow, was lined with wall cabinets and floor cupboards on both sides. The units were in birch or Douglas fir plywood and incorporated under-cabinet lighting. The provision of power points was generous, as it was in all of the houses that Wood designed, and anticipated future requirements. Those in the kitchen were set into the floor.

Figure 7.2 Kenneth Wood. Fenwycks, Camberley, Surrey. View of lower living zone
        and fireplace, 1968.
Photographer: Michael Murray. Credit: RIBA Library Photographs Collection.

A glass back door let in plenty of light and a three-quarter height, two-way
cupboard and shelf unit created a partial division between kitchen and dining zones. A pull-out ledge on the kitchen side served as a breakfast bar. As
House Beautiful observed, in what had become almost obligatory terms:
'Mrs Paton didn't want to be totally isolated from the rest of the house
when she was preparing meals.'[28] The kitchen had a drying cupboard for
clothes and was fitted with a range of modern domestic appliances, such
as a New World 72 cooker with a separate hotplate, a convenience that

Figure 7.3 Kenneth Wood. Fenwycks, Camberley. View from dining zone to upper
    level, 1968.
Credit: RIBA Library Photographs Collection.

allowed the oven to be 'fitted at whatever height above floor level that suits
the housewife best'.[29] The opening between the dining and kitchen areas
was made wider than standard and fitted with saloon-style doors to allow
hands free movement between the kitchen and dining zones (see Figures 7.3
and 7.4).[30]

The lower floor of the house was planned with future alterations in mind,
to create an additional self-contained bedroom, dressing room and bathroom from the original garage. Services for the extensions were completed

Figure 7.4 Kenneth Wood. Fenwycks, Camberley. View from dining zone to
         kitchen, 1968.
Photographer: Michael Murray. Credit: RIBA Library Photographs Collection.

as part of the original build and were selected for efficiency at this enlarged
scale. Electrical wiring was left coiled inside the walls of the initial structure,
ready to be unwound when required. Plumbing for an additional downstairs
bathroom was completed in the garage as part of the original construction.[31]
Space for a new garage, should one be required, was identified at an early
stage in the planning and the garden design evolved to allow its addition as
part of a second phase of development.

The first floor of the phase one house was also planned for alteration.
Floor-to-ceiling wall cupboards in sapele mahogany lined the corridor,

which was panelled in Douglas fir and treated with a natural seal. Clerestory lighting ran the length of the upper floor. The Patons' two young boys were given a bedroom above the garage, furthest away from the main living space. Partially divided by a large wood-panelled cupboard partition, it was arranged to allow them to share the room but have their own space within it. They appear, from contemporary articles, to have been a rather boisterous pair and the location of their bedroom, ostensibly to give them a quiet space in which to study, may well have been a blessing to other family members.

A number of interior elements were planned for future adjustment. Internal partitions and fitted cupboards in the upper corridor concealed openings designed to allow the conversion of the large master bedroom into two smaller rooms at a later date. Power sockets and wall lights were arranged to work equally well in both arrangements. Thought was also given to future requirements for independent access to the self-contained living spaces of the phase two house and provision was made within the original design. At the end of the first floor corridor a glazed opening in the exterior wall was completed as part of the phase one building ready for conversion to an additional front door with the future addition of an external staircase.

The interior of Fenwycks incorporated a variety of timbers (see Figure 7.5). The afrormosia staircase was complemented by a maple handrail and Canadian maple strip flooring in the living area. The dining room cupboard and drawer unit was faced with sapele mahogany and the dining room walls were clad with a horizontal pine strip to create a distinction from the flooring. Michael Manser commented on the results: 'The ground floor and kitchen levels are brilliantly composed with a perfect balance of timber – natural or painted – and other surfaces, like brick ... Overall, the effect is quietly opulent.'[32] He felt that the first floor was rather less successful, the abundance of timber finishes creating a rather 'sombre' feel.[33] Manser was not alone in associating Wood's use of timber with the sense of understated luxury that permeated the interiors of others houses that he designed.[34] Other materials used within the interior of Fenwycks included black clay tiles in the entrance lobby and cloakrooms and cork for its practicality and texture in the kitchen. Brightly coloured carpets from The Carpet Manufacturing Company, for which Eric Paton worked, featured throughout. Mandala, a Wilton wool carpet designed by Audrey Tanner, which won a Design of the Year award in 1959, was used in a blue in the living area and in red in the dining room and a fine blue and brown striped carpet was selected for the landing and corridor.

One of the ways in which writers situated Wood's timber-framed houses was within the context of a national history of domestic architectural development. In describing Fenwycks, Michael Manser commented on its relationship with medieval half-timber buildings: 'Despite the hundreds of years between them – and the radically modern detailing of Fenwycks – an ancestral similarity breaks through, particularly downstairs, where the brick base and timber-framing inside are most apparent.'[35] Another contemporary article, published

Figure 7.5 Kenneth Wood. Fenwycks, Camberley. View from dining zone showing
       Wood's use of timber.
Credit: Courtesy of Kenneth Wood.

in *Illustrated Carpenter and Builder*, made similar observations, identifying
Fenwycks as 'the modern development of the Tudor half-timber house'.[36] Wood
promoted historical continuities in his own writings, as a means of addressing
resistance to the use of timber construction, stating in 1961:

> As a nation we are excessively conservative in our architecture, but
> surely we cannot think that housebuilding in timber is in any way revo-
> lutionary if we remember that the use of timber as the main structural

material in a house originated and was perfected in this country three and four hundred years ago.[37]

In contrast to the historical accent of British reviews, French magazine *Votre Maison* described Fenwycks as a surprising avant-garde home, emphasising Wood's progressive use of traditional materials to create an adaptable home.[38]

After the house was completed, two extensions were made to the building in line with Wood's original plans, adding a further 2,000 square feet of accommodation. The alterations included the extension of the living/dining area by twenty-eight feet to include a new sun room and space to display the family's collection of ceramics and paintings. The kitchen was also increased in size to include a breakfast area and additional storage space. The east wing of the house was enlarged as planned, to provide self-contained accommodation with its own entrance. The re-modelled house featured in *BC Wood*, a promotional magazine published by the Council of Forest Industries of British Columbia that was aimed at a British readership. The article 'Fenwycks: a house fit for a family' showed the result of both extensions and described the benefits of planned expansion, noting:

> All too often house extensions are unsightly appendages, carried out at substantial cost and inconvenience. But, where changing needs are anticipated and provided for in the basic design of the building, adjustment and enlargement of space and facilities can be carried out easily and quickly.[39]

In its completed form, the house exemplified many of the attributes of timber construction that the Canadian lumber industries had been promoting since the early 1950s and pictures of the extended house were used in an advertisement for two educational publications by the Council of Forest Industries, *Guide to Platform Frame* and *Timber Frame Housing at its Best*.

For Wood, long-term thinking was a necessary architectural requirement and Fenwycks was envisaged as the ultimate planned environment. In contrast to the many expandable houses that appeared in British publications of the 1950s, it was designed with the full family lifecycle in mind and to meet changing needs of a particular family over time: the need for privacy and supervision at different stages in a child or young adult's life; the flexibility to bring a relative into the home and provide them with a space for independent living, or to allow adult children to continue to live at home independently if they so wished; and the chance to remain at home in old age, reducing the financial burden and burden of care involved in owning a large property by dividing it to allow the possibility of subletting.

Notes

1 There are precedents in the work of European avant-garde architects of the inter-war period. The Steel House (1926–27), designed by Georg Muche and Richard Paulick, was originally intended as an expandable house. Built on the Törten Estate at Dessau, it is now used by the Bauhaus Dessau Foundation as an information centre for the estate. See: http://www.bauhaus-dessau.de/index.php?The-steel-house-by-georg-muche-and-richard-paulick, accessed on 16 December 2012.

2 C.H. Reilly, 'The war and architecture', *Decorative Art, The Studio Yearbook, 1941* (London: The Studio Limited, 1941), 13. He appears to have been thinking, in particular, of Le Corbusier's discussion of rootedness: 'The house will no longer be an archaic entity, heavily rooted in the ground by deep foundations, built 'solid', and to which the cult of family, bloodline etc. has so long been devoted.' See, Le Corbusier, *Towards an Architecture*, with an introduction by Jean-Louis Cohen, trans. John Goodman (London: Frances Lincoln, 2008), 259.

3 See, http://www.antiquehome.org/House-Plans/1935-Sears/HomecreSthtm, accessed on 23 August 2013.

4 On *House and Garden* see Kathleen L. Endres and Therese L. Lueck, eds. *Women's Periodicals in the United States: Consumer Magazines* (Westport, CT: Greenwood Press, 1995): 150. On the Pittsburgh house see, 'Advantages of house that grows are numerous: unit will solve problem of average income family', *The Pittsburgh Press* (27 December 1936), classified section, 8.

5 See, Keith York, 'Frank Lloyd Wright's legacy in San Diego', *Save Our Heritage Organisation, Reflections Quarterly Newsletter*, 37/1 (2006). http://sohosandiego.org/reflections/2006-1/wrightlegacy.htm, accessed 16 December 2012 and San Diego Modernism Historic Context Statement, see: http://ohp.parks.ca.gov/pages/1054/files/san%20diego%20modenism%20context.pdf, accessed online on 8 July 2013.

6 The house appeared in British architectural journals in that year. See, 'House in New York', designed by Marcel Breuer', *AJ* (2 June 1949): 499–502; 'House for growing family', *A&B* (20 May 1949): 451–3.

7 'House designed by Marcel Breuer being built in museum garden', Museum of Modern Art, New York. Press Release, for release 12 January 1949. See, http://www.moma.org/learn/resources/press_archives/1940s/1949, accessed 31 August 2013.

8 'House designed by Marcel Breuer being built in museum garden.'

9 'House designed by Marcel Breuer being built in museum garden.'

10 Martin Gellen, *Accessory Apartments in Single-Family Housing* (New Brunswick, NJ: Transaction Publishers, 2012), 26–7.

11 Gellen, *Accessory Apartments*, 26–7.

12 Gellen, *Accessory Apartments*, 26–7.

13 'The lifetime home that expands and contracts', *Popular Mechanics* (October 1953): 162–3.

14 Goldfinger's sketches are reproduced in Jensen, *Modernist Semis and Terraces in England*, 209.

15 An example, by J. Kennedy Hawkes, appeared in *The Builder* in 1953. See, 'Brick Farm, Claygate, Surrey: a 3-bedroom house planned to be extended at a later date; Architect: J. Kennedy Hawkes', *The Builder* (18 December 1953): 966. Peter Barefoot, one of the British Columbia Lumber Manufacturers' Association competition panellists and Wood's travelling companion to North America, was one of those designing for expansion in this period. See 'House at Ipswich designed for the later addition of two bedrooms; Architect: Peter Barefoot', *A&BN* (20 October 1955): 490–2.

16 Taylor and Green, 'House at Kingston-on-Thames' and Stephen Gardiner 'Fitzroy Park, N.6', in F.R.S. Yorke and Penelope Whiting, *The New Small House*, 3rd enlarged edn (London: Architectural Press, 1954), 20–3, 70–1.

17 Cited in Mark Crinson, 'Picturesque and intransigent: 'creative tension' and collaboration in the early house projects of Stirling and Gowan', *Architectural History*, 50 (2007): 267–95, quotation 276.

18 'A house which grows', *House and Garden* (April 1957): 67.

19 The concept of the service core evolved in conjunction with the development of prefabricated dwellings in the inter-war period. See, Ryan E. Smith, *Prefab Architecture: A Guide to Modular Design and Construction* (Hoboken, NJ: John Wiley & Sons, 2010), 12.

20 The Parker Morris Committee was established to look at the standards of design and equipment required for family dwellings and other types of residential accommodation. The committee identified living space and heating as the two major changes required to housing, a shift in the terms in which the interior spaces of the home were understood in relation to their usability for different types of activity throughout the year. The report recognised that it was not only necessary to provide a suitable place in which a family could be together, but also rooms for independent activities that demanded 'privacy and quiet'. See, *Homes for Today and Tomorrow* (London: HMSO, 1966), 2–3.

21 *Homes for Today and Tomorrow*, 9.

22 'The furniture exhibition', *AD* (March 1961): 138. See also, 'No lack of things to live with: extremes of style at furniture show', *The Times* (2 February 1961): 8.

23 On the Ministry of Housing house see, Tatjana Schneider and Jeremy Till, *Flexible Housing* (London: Architectural Press, 2007), 73.

24 'A timber house has warmth and elasticity', *The Illustrated Carpenter and Builder* (10 March 1961): 730.

25 The house featured in: *Gas Times* (February 1961): 30–1; *Southern Service* (February 1961): 6; *CMC Look* (Spring 1961): 10–11; *Wood* (March 1961): 88–90; *The Illustrated Carpenter and Builder* (10 March 1961): 730–1; *Modern Woman* (April 1961): 83; 'Domestic lighting survey', *Light and Lighting* (June 1961): 66–175; *Home* (September 1961): 71–3; *AD* (October 1961): 453; *The Daily Mail Book of Kitchen Plans* (London: Associated Newspapers, 1964): 41; 'Planning your kitchen', *Good Housekeeping*'s Famous Series of Booklets 21 (1963): 29; *House Beautiful* (June 1963): 60–2; 'Une etonnante maison d'avant-garde', *Votre Maison* (Christmas 1962/January 1963): 56–9; BC Wood. Undated (*c.* 1972): 8–9.

26 Michael Manser, 'Halfway house', *Home* (September 1961): 71.

27 'House to fit a family', 60.

28 'House to fit a family', 60.

29 'A family house planned to grow', *Southern Service* (February 1961): 6.

30 These were appreciated by *House Beautiful*, which claimed 'No housewife has time for needless domestic work. She wants to spend her time *with* her family – not cleaning up after them'. 'House to fit a family', 60.

31 'A house in harmony', 10.

32 Manser, 'Halfway house', 71.

33 Manser, 'Halfway house', 71.

34 Ian Nairn, *Modern Buildings in Britain* (London: London Transport, 1964), 105.

35 Manser, 'Halfway house', 71.

36 'A timber house has warmth and elasticity', *Illustrated Carpenter and Builder* (10 March 1961): 730.

37 'Building homes of timber', *Financial Times* (13 November 1961): 40. The article, written by Wood, appeared under the by-line 'a correspondent'.

38 'Une etonnante maison d'avant-garde', *Votre Maison* (Christmas 1962/January 1963): 56–9.

39 BC Wood, ed. D.W. Macklin (London: Council of the Forest Industries of British Columbia, undated, [*c.* 1972]), 8. KWPP.

8 Show house
Hampton House, 1961

In the same year that Vincent House was selected for inclusion in the Architecture Today exhibition, Hampton House at Hampton (1961), designed by Wood to promote the work of a newly formed housing development company, Hampton House Developments Limited, was named as *Ideal Home* magazine's House of the Year (see Figure 8.1).[1] The project was one in which Wood's human-centred approach to the organisation of domestic space was, due to the commercial structure of the project, somewhat compromised in favour of his client's vision for the modern home – a vision derived quite directly from European avant-garde thinking of the inter-war period. As an example of a collaboration between an architect, a housing development company and a popular home magazine, Hampton House offers a window on the modern middle-class home of the early 1960s in ideal form, and provides a vantage point from which to consider the professional structures and intersecting consumer practices that informed its design and representation.

Hampton House Developments was a joint venture between Norman Barker, a businessman with a keen entrepreneurial spirit, and Charlotte Parker, a potter and interior designer who had a flair for marketing and good media contacts.[2] Parker was the driving force behind the project. Her family roots were in Europe and at the time that Barker met her, in the spring of 1960, she was selling her ceramic designs through Heals department store.[3] She had, prior to his involvement in the scheme, already begun to formulate her plans for a modern, flexibly designed house for middle-class professional clients and had found and purchased a plot of land on which to test its commercial viability.[4] Credited in the development company's publicity material as Building Director, Barker had no professional experience in the fields of design or construction. The prototype house was not the financial success that the couple had hoped for and the venture failed, leaving Barker to go on to establish other businesses and Parker to join Bentalls department store at Kingston.[5]

Having seen his BCLMA competition house (Chapter 5), it was Parker who first contacted Wood to seek his involvement with her new business.[6] His main input into the project was at the planning and architectural design stage. Parker, as this chapter will show, had a clear vision for the house and

Figure 8.1 Kenneth Wood. Hampton House, Hampton. From: *Ideal Home Magazine*,
June 1961.
Credit: Courtesy of IPC Media.

the type of living space that she believed would appeal to contemporary
buyers. She was instrumental in the conceptual development of the design
and was responsible for the decoration and furnishing of the interior.

The prototype house, a detached single-family dwelling, was built on a
quarter of an acre of land that had once formed part of the garden of a
large private house, Berkeley House, on Upper Sunbury Road, close to the
River Thames at Hampton. Wood designed the house with a load-bearing,
L-shaped ground floor in Ibstock brick, with structural steel pilotis supporting a projecting timber-framed first floor. All of the timber for the project
was supplied by Montague L. Meyer Limited and was prepared in a temporary workshop above a bank at nearby East Molesey, an arrangement
that speaks to the somewhat provisional nature of the whole project.[7] Wood
sited the house to the centre of the plot and turned it away from Berkeley
House to give the main living rooms on the first floor a south-west orientation and views over an adjacent sports ground. To one side of the plot a

separate outbuilding accommodated a double garage, a garden store, and a ceramics workshop for Parker, whose work was incorporated into the interior in a large fire surround in the main living area. The rationale for including a personal studio space within this commercial project is unclear. Parker perhaps saw it as an opportunity to promote her work – the inclusion of the fireplace certainly suggests so – but might also indicate that she found it difficult to separate her personal aspirations from the more generalised needs of potential clients. This is to some extent borne out in the way that she and Barker approached the furnishing and decoration of the Hampton House interior.

As Parker and Barker hoped to use the house as a platform for larger housing schemes, Wood designed the house for series construction in clusters or groups of staggered terraces in relatively high-density configurations. Although historically associated with standard housing types, the success of Span's modern terraced housing had begun to change perceptions within the building industry and by the early 1960s house-building firms were starting to reconsider the terrace as an appropriate form for luxury houses. Wood was well aware of this development. In the same year that Hampton House appeared in *Ideal Home* magazine, *Woman's Journal* promoted is own House of the Year, the Echelon House. Part of a luxury development on Copse Hill at Wimbledon, it was similarly designed for series construction in a staggered formation.[8] In *Modern Buildings in London*, published three years after its completion, Ian Nairn described the Copse Hill estate as 'a good example of how one of the biggest building firms in the country is now prepared to provide modern houses in sensitively landscaped surroundings' and identified the source of this new thinking:

> The person responsible is of course Eric Lyons with his Span housing; however, this estate is not merely imitating Span, but has its own crisper flavour. Two-storey stepped terraces near the main road, which are as good as can be … This is still, so far, a London phenomenon, and what such firms do in the provinces is nowhere near this standard.[9]

Woman's Journal also recognised, albeit implicitly, the influence of Span on the pattern of suburban housing:

> In the suburbs, estate developments of 'off-the-peg' terraced housing with communal landscaping are becoming increasingly popular, and the idea that respectability can only be attained with detached property is rapidly changing. With this change, comes the idea that new houses can be terraced and yet be in the luxury class.[10]

Rapid construction was a major concern for Parker and Barker, who recognised that their future success would depend on their ability to complete projects at speed. Hampton House was used to trial a technique for all-weather building, which involved shrouding the site under a large, fan-

Figure 8.2 Kenneth Wood. Hampton House, Hampton, showing the tent for all-weather construction in 1960. From: *Ideal Home Magazine*, June 1961.
Credit: Courtesy of IPC Media.

inflated tent made from waterproof coated nylon (see Figure 8.2). The tent was supplied with electricity to maximise winter working hours, which permitted the protected construction of the house to first floor height. It is unclear where this idea originated. Wood may well have been alerted to the practice during his study visit to Canada. In the late 1950s, Visking (part of Union Carbide) marketed an inflatable 'hot-air bubble' under the name Visqueen, which it promoted to the Canadian construction industry as a means of allowing concrete to be poured in 'sub-zero weather'.[11] The process of setting out and inflating the tent was recorded by British Pathé for a news item that reported on Hampton House as the first British example of a house to be built using a tent of this type. It also featured in the local newspaper, the *Surrey Comet*, bringing welcome advance local publicity to the project.[12] Completed between October 1960 and February 1961, Hampton House opened to the public as a show home in late May 1961 and remained open until the middle of November of that year.

The development of Hampton House was somewhat unorthodox from the start. In 1960, with basic financing in place, Parker approached the editor of *Ideal Home* magazine with the idea of gaining exposure for the project by offering it to the magazine to promote modern design and the concept of flexible living. Following initial discussions, *Ideal Home* agreed to feature Hampton House as its House of the Year in the following year. Parker and Barker used a letter from *Ideal Home*, which guaranteed future publicity for the project, to secure the additional finance that they required to complete it.[13] In addition to an extensive article in the magazine's June 1961 issue, Hampton House was promoted to the public through an illustrated thirty-two-page brochure that was compiled by Parker and produced by her father's printing company. The brochure contained comprehensive information on the design of the house and advised potential clients that 'designs for other houses on this principle' were already in development.[14]

The main feature of the house that Parker wished to promote was a movable partition system that was intended to allow the effortless alteration of its interior spaces according to day-to-day needs.[15] In its in-house journal, the De La Rue Company, which helped to produce the system, emphasised the spatial dimensions of Parker's role as an interior designer as well as its own contribution to the enterprise:

> As befits a specialist in planning conversions of small spaces, Mrs Parker has designed a house in which there are no spare rooms. Quite apart from aesthetic considerations, the Hampton House is very much a product of the period in which rising costs have seen many people being forced into houses with smaller living spaces. This together with its use of an excellent mixture of plastics and non-plastics suggest that 1961's Ideal Home of the year may be taken as a guide for designers and architects for many years to come.[16]

In plan, Hampton House looks back to the experiments of European avant-garde architects of the inter-war period and to a particular model of spatial flexibility, expressed in the interior design of the pioneering Rietveld Schröder House (1924) in Utrecht, commissioned by Truus Schröder-Schräder and designed by Gerrit Thomas Rietveld, and the somewhat later 2 Willow Road (1939), designed by Ernö Goldfinger as his family home in London (see Figure 8.3).

The principle of spatial flexibility influenced the development of a wide range of public and private interiors after the Second World War and is evident in the vast array of folding and movable wall products that were advertised in British architectural journals in the 1950s, among them 'Modernfold' doors produced by Home Fittings (Gt Britain) Limited, Holoplast Limited's 'Holoplast Movable Walls' panel system, and sliding doors produced by The British Trolley Track Company Ltd.[17] While Wood was totally committed to the principle of flexibility, he was pursuing quite a different path of

Figure 8.3 Ernö Goldfinger. The nursery at 2 Willow Road, Hampstead, London, 1939.
Photographers: Dell and Wainwright. Credit: Architectural Press Archive / RIBA
Library Photographs Collection.

development in his work for private clients, favouring open plans with clear
circulation routes over the type of user-determined, folding and unfolding
spaces of Parker's imagination. The result of their collaboration was a somewhat compromised plan that aimed to be user-determined but was, in its
realised form, somewhat prescriptive and in several respects failed to meet
the standard set by Wood in other projects.

The ground floor of Hampton House was arranged as an entrance hall
with built-in storage and an open-tread staircase to the first floor. To the
left of the hall were a utility room, 'a place for useful and necessary storage
and activities' that provided 'hanging and drying space for wet and outdoor
clothing', a shower room that functioned as 'a second bathroom for guests,
wet children and sailing enthusiasts alike' and a cupboard that housed a
warm-air heating system by Johnson and Starley, which was imported from
Canada.[18] To the right of the hall was a single room that was designed for use
as study and for subdivision by means of the partitioning system, the benefits
of which were outlined in the Hampton House Promotional Brochure:

> The study is a typical example of the advantage of flexibility. When the
> family is in sole occupation the whole of the space is in use as study.
> When guests arrive to stay, there is no need to give up a whole room;
> each portion can be used indefinitely as a self-contained unit, one as a
> bedroom, the other as a study. Similarly, this space could be a large day

Figure 8.4 Kenneth Wood. Hampton House, Hampton. The living room configured for daytime use. From: *Ideal Home* Magazine, June 1961.
Credit: Courtesy of IPC Media.

nursery and playroom partitioned at night to give individual bedrooms. The spare bedroom afforded by partitioning the study has the further advantage of providing a ground floor sickroom in case of need.[19]

Outside, a covered terrace provided shade in the summer and a protected play area for children in bad weather (see Figure. 8.1). A small goods lift allowed the transport of food from the first floor kitchen to the terrace, a consideration of outdoor dining requirements that reflects the place of technology in shaping modern indoor/outdoor lifestyles, for which precedents can be found in British inter-war houses such as Maxwell Fry's Miramonte.[20] An article on Hampton House that appeared in *The Times* newspaper criticised the terrace, noting that because it was overhung by the first floor it not only received little sun, but also restricted light to part of the ground floor.[21] Long-term owners of the house, who raised their family there from the late 1960s, found it too dark to be useful and integrated it into the interior as kitchen-dining room in the mid 1970s.[22]

The first floor of Hampton House was planned as a large L-shaped living area, 39 feet in length, extending the width of the front of the building. To the rear were smaller, fixed spaces: a kitchen, a family bathroom and the

Figure 8.5 Kenneth Wood. Hampton House, Hampton. The living room configured
   as a guest bedroom. From: *Ideal Home* Magazine, June 1961.
Credit: Courtesy of IPC Media.

only permanent bedroom with solid walls. The partitioning scheme for the
first floor was extensive. Fourteen movable panels were designed to allow
the division of the living area into two bedrooms and a separate living room
and dining room (see Figures 8.4 and 8.5). As Parker explained:

> The system allows for the needs of ever-changing family life and affords
> the advantages of both fixed and open planning. We have chosen furni-
> ture and fittings which are versatile in use and help to demonstrate the
> possible changes in the house.[23]

Although the cost of construction was relatively high, the Hampton House
brochure emphasised the long-term economy of the design: 'By virtue of its
flexibility it is an extremely inexpensive house to furnish, and although the
various fittings and systems add to the initial expense, they make for very
low running costs and minimum maintenance.'[24]

Parker's original plan for an interior with a series of lightweight sliding partitions was never realised. The construction of the panels was complex and involved a number of companies. Monsanto Chemicals Limited

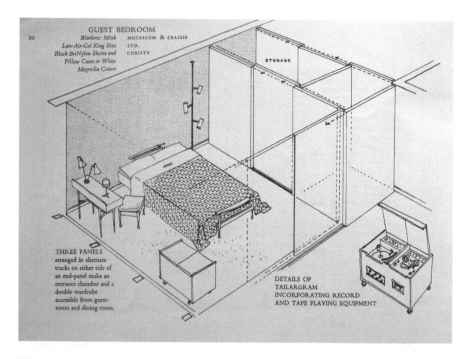

GUEST BEDROOM
20   *Blankets: Mink*   MCCALLUM & CRAIGIE
Lan-Air-Cel King Size   LTD.
Black BriNylon Sheets and   CHRISTY
Pillow Cases or White
Magnolia Cotton

STORAGE

THREE PANELS
arranged in alternate
tracks on either side of
an end-panel make an
entrance chamber and a
double wardrobe
accessible from guest-
room and dining room.

DETAILS OF
TAILARGRAM
INCORPORATING RECORD
AND TAPE PLAYING EQUIPMENT

Figure 8.6 Kenneth Wood. Hampton House, 1960. Details of the partition system. Credit: Courtesy of Kenneth Wood.

supplied fire retardant polystyrene beads that were made into boards by Levecto Limited. Edging tracks, nylon tubes, bolts and ferrules were supplied by Alcan Industries, Polypenco, and Chubb & Sons. The decorative surfaces of the panels were applied by Modern Veneering Limited. The completed panels had to fit into a handcrafted timber frame and problems with the accuracy of the joinery led to the failure of the track-running system into which they were designed to fit.[25] The compromise – a radical rethinking of Parker's original plan – was the installation of a demountable system of seven and a half foot high wall panels that could be locked into position or stored at the end of a room when they were not in use (see Figure 8.6).

Although *Ideal Home* commented that 'weighing only 18 lb. each, the panels could be lifted by a woman' other contemporary sources suggest that they required some effort to reconfigure.[26] A 'special correspondent' for *The Times* newspaper appreciated the concept but not its execution:

> While the idea is good and the method ingenious, the real result is to give a prefabricated air to an expensive purchase. Prices, exclusive of the cost of the land, range from £9,000 to £7,000 for an economy model. A

cut-down version with less space and fewer fittings would cost £5,300 plus land. The utility feeling comes primarily from the use of the panels themselves; they have obvious joints and are no more soundproof than most partitions. Metal grooves like miniature tramlines in the floor do nothing to help. They can be partly disguised by carpets, but only if separate strips of carpet are used so that the grooves are still accessible. At present there are also grooves in the ceiling, but it is planned to hide these in future models.[27]

Ideal Home suggested that the panels looked 'very little different from any normal wall, only the lines between panels and the lifting finger-slots just being visible'.[28]

Contemporary coverage of the house hints at Parker's frustration with the end result, which fell short of the seamless modernity to which she aspired.

> It takes 'about half an hour and a lot of brawn' to add or subtract a room, according to Mrs Charlotte Parker, the interior design consultant who had the idea of flexible rooms and lives in this prototype. 'You wouldn't do it for fun', she added. 'You wouldn't do it for friends if they were staying only for one night.[29]

Instead Parker suggested that it was an appropriate design for 'those whose families are likely to change in size, or who like having people to stay but, when alone, would rather utilize all space than keep permanent spare rooms'.[30] Parker's defence appears somewhat half-hearted and the panel system was rather short-lived and had been removed from Hampton House by 1967.

Apparent from *Ideal Home* magazine's colour photographs of the house is Parker's striking decorative scheme for the interior. A riot of colours and textures greeted visitors in the hallway: black Vinylex tiles by Semtex Limited, vertical striped wallpaper (Tapio 8705) by John Line and Sons, a contrasting heavily patterned curtain in a textile by Finnish Designs and an alternating purple, royal blue and indigo Shildon BriNylon carpet by Alfred Morris Limited on the stair treads, which Wood had designed in a tropical African hardwood, with the clear intention that they be left uncovered.[31] The wall panels were decorated with washable Formica or wallpaper. *Ideal Home* described the partition system as 'revolutionary' and noted that it was in no way restrictive when it came to decoration, as it allowed the division of the interior into different rooms, 'each with its own colour scheme'.[32] The Formica-surfaced panels had 'tweed' or 'mosaic' patterns on one side and 'polar white' on the reverse. *Ideal Home* showed the dining area with the 'polar white' arrangement, complemented by blue, white and yellow striped curtains and blue-upholstered dining chairs. The living area, with a mustard yellow colour scheme, was shown in use with a day bed at one end and reconfigured as a bedroom with a pink and white colour scheme. The furniture for the house was selected to work within these different room settings. As *Ideal Home* magazine noted of the bedroom layout: 'Twin divans that formerly looked like an elegant L-shaped wall sofa now stand free to show themselves as beds with rose-print covers, and a neat writing desk transforms itself into a graceful dressing table.'[33] Collectively, the panels, the opportunities that they offered for varying the decor in line with social use, and the furnishing of the interior so that 'everything that possibly can has a second use' expressed Parker's vision of interior flexibility.[34]

Hampton House was a highly commercialised project from start to finish. In addition to plans and drawings, the promotional brochure for Hampton House Developments itemised all of the fittings, furniture, loose furnishings and decorative objects that were displayed within it, along with the name of each supplier. Firms donated goods in the knowledge that whatever they gave would not only be seen by visitors to the show house, but would also benefit from editorial coverage in *Ideal Home* magazine. Some, according to Barker, also paid for the privilege.[35] Although the promotional brochure indicated that the house was furnished with the support of Heals department store, with which Parker had connections, the couple approached multiple suppliers in completing the interior. Placing themselves in the arena of the private consumer and adopting what were effectively amateur practices of homemaking, they made extensive use of popular magazines to select and source furniture and appliances, flicking through and choosing things that they liked before contacting companies to see whether they would be willing to donate products. The resulting interiors appear less studied than similar examples of exhibition or show homes completed as a piece by a single commercial contractor.

Furniture selected for the living area included award-winning String wall units, designed in 1949 by Swedish architect Nils Strinning, and supplied by Swedia Imports Limited; several items of furniture supplied by Guy Rogers Limited, including a rocking chair, a Manhattan studio couch in Tibor Brecon Olive deep texture and three Camino chairs covered in Tibor 'Ancona' Persimmon; and a Heals dining table with chairs covered in Tibor Avon purple fabric.[36] The house was dressed with a range of household objects that included plant troughs with slatted magazine shelves by Peter Cuddon, a white china cat by Rosenthal, desk accessories by WHSmith, a noiseless hairdryer and stand by Murphy, an autorise clock by Smiths Clocks and Watches, a gas table lighter by Ronson, gramophone records by EMI Records, a Sparklets soda syphon by The British Oxygen Company, and a model 65 cleaner by Electrolux. A house of brands, its cupboards were filled with everyday essentials – Andrex, Pat Washing Powder, Fairy Liquid, Boots' Home First Aid Kit, Fairy Snow, Pink Camay, Stergene, Gleem, Ajax, Scot Towels, Durazone-Choice Aerosols and Disinfectants, Whitbread Beer, Schweppes Drinks, Nestea, Nescafe, Lyons Maid Ice Cream, Mary Baker Cake Mix and Hovis.

Aspects of the architectural design of Hampton House suggest that Wood's experience in domestic planning was undervalued by his clients and again seems to point to Parker's reliance on earlier domestic paradigms in her approach to the design of the interior. This is most evident in the planning of the space-saving kitchen, which Parker described as 'a laboratory for one woman' and listed as one of the top three features of Hampton House, along with the flexibility of the panel system and the adaptable furniture that she had selected to work with it.[37] The kitchen was lit from above and had no outlook, an aspect of the design that provoked negative comment from one reviewer, who acknowledged that it was 'well equipped with labour-saving devices' but criticised its lack of view.[38] The experience of the main female occupant of the house, who took up residence with her young family in the late 1960s, lends support to this criticism. She has recounted that its impracticality for family life affected her enjoyment of her home when her children were young. She often had to pin back the kitchen door to observe the playpen in the living area and was unable to see the garden to supervise her children at play when she was working in the kitchen.[39] Neither did the kitchen satisfy her psychological requirements. She felt confined by the space and remembers one of her young daughters drawing her a picture of a window with a garden view to put in front of the sink to brighten up the blank wall.[40]

Although Wood designed a number of galley kitchens for smaller houses, these compact working environments were planned in consideration of a variety of associated household requirements – refuse disposal, the washing and drying of laundry, the location of the front door, the observation of children, the cultivation of the garden and the service of food. They were also of more open design, connected to the spaces around them. Those that he designed for larger houses were conceived as social environments, with adjoining breakfast spaces or counters for informal dining in line with changing middle-class attitudes at the time. A rather different model of kitchen to that of the 'laboratory for one woman' was chosen for Wimpey's Echelon House of the same date and suggests that Parker was somewhat out of step with the market. *Woman's Journal* described it:

> The desire to plan a labour-saving kitchen in our own homes only too often produces a shiny, clinical appearance, more like a laboratory than part of a home; and we have tried to avoid this in the House of the Year kitchen without any sacrifice of efficiency.[41]

Other design decisions were made in response to *Ideal Home* magazine's production deadlines. The garden was completed at speed to provide suitable exterior photographs for the magazine.[42] Here, too, is little evidence of the thoughtful approach to the design of exterior spaces that characterised Wood's earlier projects. Although the exterior landscaping was designed by a leading figure in the world of garden design, Paul Temple, whose company was based at Hampton, it was his skill in designing photogenic instant gardens and displays for trade shows and exhibitions that was called upon in this project.[43] The grounds were mainly set to lawn, with a small paved area to the front of the covered terrace, for which Temple created three raised flower beds and a small rockery alongside the driveway.

Hampton House changed ownership several times in the early 1960s, perhaps suggesting that occupants found it difficult to accommodate themselves to the space. When it came up for sale in 1967, Bentalls department store's Estate Offices described it as a 'unique house' in the 'contemporary style' with a 'simple and practical plan' that was 'ideal for those with a growing family' and listed its special features: warm-air central heating, ample electric points, double-glazed windows, concealed plumbing, built-in wardrobes, cupboards and units and a 'luxurious and spacious patio with marble flooring intended for summer living'.[44] Sales particulars are often helpful in identifying those factors with most relevance or appeal to contemporary consumers in making their purchasing decisions. Those for Hampton House strike a balance between what were, by then, widely understood as essential requirements for modern living – storage, technology and a flexible plan – and the luxury materials and modern aesthetic that set it apart from the majority of other family houses in the area.[45]

The commercial structure of the Hampton House scheme and lack of expertise of its developers made it an unusual project for Wood. Although his hand is clear, Parker's stamp on the design was such that Hampton House is different in character to the houses that he designed for private clients. Its roots are firmly planted in inter-war Europe, in designs that arose in response to the contemporary needs of that period. The partitioning system and rational kitchen design, in particular, look back to earlier models of domestic design and expressions of interior flexibility. Parker's furnishing and decoration of the interior and her use of fashionable colour and pattern to create multiple interior identities conflicted with Wood's commitment to the principle that spaces should derive their character architecturally, through the process of inhabitation, and from the personal possessions of their owners.[46] While the poor execution of the interior partition system was undoubtedly a major factor in the failure of the house to live up to initial expectations, it was perhaps inevitable that the result would be compromised given the conflicting values that Wood and Parker brought to the project.

Notes

1 *Ideal Home* (June 1961): 49–60.
2 Norman Barker, in discussion with the author, March 2012.
3 Her parents, Austrian Jews, had immigrated and settled in London where her father established a printing business, Edwards & Brune Ltd. Norman Barker, in discussion with the author, March 2012. The Hampton House marketing brochure lists her qualifications as BA, MIPI. Hampton House Promotional Brochure, KWPP.

4 Parker was recently divorced and used her divorce settlement to develop the project. Norman Barker, in discussion with the author, March 2012.

5 Norman Barker, in discussion with the author, March 2012.

6 Norman Barker, in discussion with the author, March 2012.

7 Norman Barker, in discussion with the author, March 2012.

8 The Echelon House was designed by George Wimpey's in-house architects, E.V. Collins, H.E. Forman and A.B. Ruggeri and was priced at £10,750. See 'House of the Year', *Woman's Journal* (March 1961). It was one of several houses of 1961 that appeared in magazines and newspapers. Another notable example is Peter Womersley's House of the Year for the *Sunday Times*, a timber-framed family house with a first floor living area that opened onto a large terrace. The interior planning and furnishing were by Gordon Good, a recent graduate from the School of Interior Design at the Royal College of Art, with Margaret Casson and staff of the *Sunday Times* women's pages. The house featured in *Sunday Times* articles on 30 July, 6 August and 13 August 1961.

9 Ian Nairn, *Modern Buildings in London* (London: London Transport, 1964), 105.

10 'House of the Year', *Woman's Journal* (March 1961): 14.

11 Advertisement. Pour concrete in sub-zero weather inside a Visqueen polyethylene film 'Hot-Air Bubble' in *The Canadian Architect* (February 1959): 75.

12 Film of the construction of the 'Air House', as British Pathé christened it, can be seen at: http://www.britishpathe.com/record.php?id=378. Film date 28 November 1960, canister CP30, tape PM0119, film 119.17. See also, 'Rain won't bother them now', *Surrey Comet* (5 November 1960): 12.

13 Norman Barker, in discussion with the author, March 2012.

14 Hampton House Promotional Brochure, 2. KWPP.

15 Hampton House Promotional Brochure, 5. KWPP.

16 'Beyond the fringe', *De La Rue Journal* 42 (Spring 1962): 42–5, on 45.

17 Home Fittings (Gt Britain) Limited's 'Modernfold' doors were advertised with images of the division of a large room into two spaces with the copy 'a new view of the space problem' and 'make more room with Modernfold expanding walls and doors', see *AJ* (19 January 1956): lxviii. The British Trolley Track company advertised its 'Coburn System' sliding doors as 'space-saving, convenient, permanently trouble-free and inexpensive' in *AJ* (19 January 1956): lxxv. Holoplast advertised the streamlined production and simplified design of its movable walls with an images of commercial interior, see *A&B* (26 March 1958): 1. In October 1957, Compactum Limited opened its 'Partitioning Centre' where architects and clients could see a range of demountable partitioning schemes for use in offices. See examples of advertisements in *AJ* Supplement (16 January 1958): 38 and *A&B* (26 March 1958): 13.

18 Hampton House Promotional Brochure, 6. KWPP.

19 Hampton House Promotional Brochure, 5. KWPP.

20 Miramonte had a lift to allow the service of food to its roof terrace. See, 'A Surrey house in a park', *AR*, (November 1937): 187–92.

21 'Rooms which you can add or subtract', *The Times* (24 May 1961): 17.

22 Sylvia Disley, in discussion with the author, March 2011.

23 Hampton House Promotional Brochure, 6. KWPP.

24 Hampton House Promotional Brochure, 3. KWPP.

25 Barker also recalls a problem with one of the timber deliveries being left out in the rain. Norman Barker, in discussion with the author, March 2012.

26 Ellis, 'House of the Year', 52.

27 'Rooms which you can add or subtract', 17. The house also received some international coverage, appearing in a short illustrated article in the Modern Homes Supplement of *The Singapore Free Press* in June 1961, in an article that described its partitioning system

An ingenious system of movable panels gives a remarkable flexibility to the interior. The large L-shaped living area can be partitioned to create as many as six separate rooms, each with its own colour scheme. This means that the interior can be altered to meet the changing requirements of a growing or growing-up family.

'Flexible 'House of the Year', *The Singapore Free Press*, Modern Homes Supplement (29 June 1961): 12.

28 Ellis, 'House of the Year', 52.
29 'Rooms which you can add or subtract', 17.
30 'Rooms which you can add or subtract', 17.
31 Ellis, 'House of the Year', 49.
32 Ellis, 'House of the Year', 52.
33 Ellis, 'House of the Year', 54.
34 Ellis, 'House of the Year', 54.
35 Norman Barker, in discussion with the author, March 2012.
36 On Tibor Reich see, M.A. Hann and K. Powers, 'Tibor Reich: a textile designer working in Stratford', *Textile History* 40/2 (November 2009): 212–28.
37 'Beyond the fringe', 43.
38 'Rooms which you can add or subtract', 17.
39 Sylvia Disley, in discussion with the author, March 2011.
40 Sylvia Disley, in discussion with the author, March 2011.
41 'House of the Year', *Woman's Journal*, 14.
42 Norman Barker, in discussion with the author, March 2012.
43 On Paul Temple, see his obituary in the *Telegraph*, http://www.telegraph.co.uk/news/obituaries/1543652/Paul-Temple.html, accessed on 29 November 2012.
44 Hampton House, housing particulars, Bentalls Estate Offices, Wood Street, Kingston, 1967, DPP.
45 Reference was also made to the possibility of reinstating the partition system. Hampton House, housing particulars, Bentalls Estate Offices, Wood Street, Kingston, 1967, DPP.
46 As Wells Coates had observed in relation to the design of modern dwellings in the 1930s:

> Our *real* possessions – that give personality and character to our dwellings and the life we lead – are, let us say to start off with: good manners; and good taste in clothes; our crockery, glassware and other small objects of personal choice or vanity; our own books, pictures, sculpture, musical instruments, even the quality of our wireless sets.

Wells Coates from the 24 May 1933 broadcast, 'Modern dwellings for modern needs' in *Documents: A Collection of Source Material on the Modern Movement* (Milton Keynes: The Open University Press, 1979), 73.

9 Converted house
Torrent House, 1965

Torrent House at East Molesey in Surrey was designed for a graphic designer, Peter Devenish, and his family (see Figure 9.1).[1] An imaginative repurposing of a small industrial building, the house was created from a redundant London Passenger Transport Board electricity substation that had been used to power a local trolley bus service until 1962. Like Whitewood, Torrent House was conceived as a studio house for family occupation, but its design reflects a different set of aspirations for the relationship between work and family life. Completed in 1965, the house received a Civic Trust award in the same year.

The national housing shortage that followed the Second World War stimulated an interest in the adaptation of disused buildings for domestic use. During the years in which restrictions on private development remained in place, planning authorities often looked favourably on private initiatives to help increase the housing stock. One example is a small house at Woldingham in Surrey, which featured in *House and Garden* magazine in 1949. Designed by Elie Mayorcas, it was created from a garden hut that its owners had been given permission to convert to free up their large house for family occupation.[2]

In 1954, just as restrictions on private building were coming to an end and in response to a perceived need for guidance on projects of this type, *Country Life* magazine published *New Homes from Old Buildings*.[3] Written by H. Dalton Clifford and R.E. Enthoven, the book was aimed at a general readership. Along with advice on the practicalities of undertaking conversions at a time of labour and material shortages, its authors offered guidance on aesthetic matters. While claiming no ambition to educate their readers, Dalton Clifford and Enthoven's text was instructive in its discussion of the visual considerations involved in converting older buildings, promoting the need for clarity between old and new elements and advising readers to make their additions in an authentic contemporary manner.[4]

The title *New Homes from Old Buildings* suggested a breadth that was only partially reflected in the book's content. It was more a case of new homes from old houses, as most of the projects described involved the adaptation of existing residential buildings: country houses, cottages, town houses, mews buildings and urban terraces. Among a handful of non-

Figure 9.1 Kenneth Wood. Torrent House, Hampton Court, from a leaflet produced
by Peter Devenish.
Credit: Courtesy of Kenneth Wood.

domestic examples were the conversion of a seventeenth-century water mill
in Hertfordshire by Dennis Lennon and the alteration of a Kentish oast
house by F.L. Marcus and Trevor Dannatt. Discussion of the water mill
centred on Lennon's success in creating a modern dwelling from a cluster of older buildings. The authors called attention to Lennon's modern
sensibility:

> Because the conversion has been carried out with a genuinely contem-
> porary feeling, it blends perfectly with the original buildings which were
> equally 'contemporary' in their day. Genuine works of different periods

always fit well together; it is only when the genuine comes in contact with the bogus that a clash occurs.[5]

Their review of Marcus and Dannatt's oast house conversion centred on the realisation of a modern plan from a building of functional design: 'In designing the alterations the spaciousness of the original building has been preserved. The planning is open, with a marked absence of narrow passages and pokey corners. Glazed doors have been used to give long vistas.'[6] The authors appreciated the way in which the architects had preserved the character of the early twentieth-century buildings to create a new dwelling from was 'still unmistakably, a group of Kentish oast-houses'.[7] The question of local style was also examined in relation to the conversion of a boat house at Dartmouth, by E.F. Tew, that lent itself to discussion of the need for a reasoned approach to design:

> It is a pity that the Victorian character of the original building should have been completely lost in the process of conversion, but the old windows had to be scrapped since their sills were at floor level, and it would not have been logical to replace them with new ones of Gothic shape.[8]

The introduction of the Civic Amenities Act of 1967, 'to make further provision for the protection and improvement of buildings of architectural or historic interest and of the character of areas of such interest' stimulated a more broadly based consideration of the reuse of disused buildings which gathered momentum in response to the economic downturn of the 1970s.[9] Sherban Cantacuzino's *New Uses For Old Buildings* (1975) is one of several books on the subject that adopted a wider framework and began to give greater thought to industrial buildings and their suitability for conversion to other uses.[10] In it Cantacuzino considered strategies for reusing redundant structures such as pumping stations, mills, warehouses and religious buildings and in addition to their adaptation for domestic purposes examined their reuse as university buildings, cultural centres, offices and museums, among others.[11] The book included national and international examples, selected for their 'visual importance in the urban or rural context, for the social or cultural importance of their new use and for the quality of their design, having regard to the character of the original building'.[12]

Torrent House was completed in 1965, at roughly the midpoint between the two publications. Shaped in response to the conditions that had stimulated an interest in domestic repurposing in the 1950s, it also reflects an emerging interest in the reuse of industrial buildings that gathered pace in the following decade. In his own positioning of the project, Wood emphasised its originality and distinctiveness from the general pattern of domestic conversions at the time. Contemporary articles on Torrent House also promoted the novelty of the project, focusing on its appearance in relation to its riverside setting, the historic built environment in which it was located, and the 'subtopian' character of the industrial building from which it was created.

Figure 9.2 Kenneth Wood. Torrent House, Hampton Court. The disused substation structure before conversion.
Credit: Courtesy of Kenneth Wood.

The substation was situated opposite Ash Island on the River Thames, just downstream from Molesey Weir, close the historic site of Hampton Court Palace, and on the opposite bank of the river to Wood's home and office (see Figure 9.2). The single-storey building occupied a narrow plot and had a flat, split-level roof that was higher toward the road to the front of the plot and lower to the riverside. Having observed that it was no longer operational, Wood was attracted by the potential of the attractive location and the possibility of developing the site for residential use. Although he had no client in mind, he made enquiries about the land and began to evolve his ideas for a single-family house on the plot. He soon realised that demolition of the brick and concrete substation would be costly and set about working with the existing building to create a modern house suitable for its riverside setting. Soon after the site was marketed he was approached by Peter Devenish, a keen sailor who was looking for a plot on which to build and hoped to find something affordable and close to the river. Wood introduced Devenish to the site, which he went on to acquire at a favourable price due to the off-putting presence of the substation and the proximity of neighbouring buildings, which placed financial and practical limitations on its development.

The project presented an opportunity for the type inventiveness that Wood most enjoyed and was the first major project that his assistant, Martin Warne, was given to manage. The narrowness of the plot afforded limited scope for

Figure 9.3 Kenneth Wood. Torrent House, Hampton Court. The split-level living area.
Photographer: Michael Wickham. Credit: Courtesy of Kenneth Wood.

extending beyond the footprint of the substation building. To minimise the
cost of construction Wood avoided major alterations to its heavy structure,
adding a highly insulated, lightweight upper floor, cantilevered to the river
side to maximise interior space (see Figure 9.3). For reasons of privacy the east
and west walls were left relatively free of openings, apart from the first floor
kitchen, which had good views toward Molesey Weir. The variation in the
substation roof level was used to define two zones within the main first floor
volume, an upper dining area and a lower living space. Large sliding windows
opened the interior to the river, giving both living areas unobstructed views.

Echoing the colours of nearby boathouses, the design of the exterior of Torrent House made visual reference to its riverside location. The brickwork on the ground floor was painted black to conceal the use of new materials around the altered ground floor window openings and the cantilevered first floor addition was clad in silvery grey asbestos sheets. The original cladding, selected for reasons of fire resistance, has since been replaced with vertical timber board, somewhat obscuring the visual allusion of Wood's original design.

The original substation structure was arranged to accommodate two bedrooms and a small suite of commercial rooms – a reception area, studio and storage room. The studio extended the width of the building on the river side and two bedrooms were located to the front of the house, below two further bedrooms on the first floor. Wood's unusual window designs for the front elevation were devised to provide adequate light to the bedrooms while minimising traffic noise from the road outside. The provision of a ground floor cloakroom allowed the studio to function as a self-contained working space during the daytime, when the ground floor bedrooms were not in use. A spiral staircase in metal and afrormosia – chosen because it required limited piercing of the solid concrete structure and was more efficient in its use of space than a standard staircase – linked the two floors. A hexagonal lantern lit the staircase from above and brought additional natural light into the upper dining area. One contemporary writer suggested that it gave 'the visitor the impression of being almost in a lunar observatory' (see Figure 9.4).[13] The interior was plainly furnished and finished quite simply, with white painted walls, a wooden floor in Tasmanian oak strip and a ceiling in fir boarding. Early photographs show musical instruments hung on the walls and a large pampas grass in an oversized floor vase casting dramatic shadows across the interior. The exterior spaces of the house, which Wood planned for later completion by the owners, included a carport to one side and landscaping to the front, with a walled courtyard to act as a sound buffer and obscure the busy main road. He also devised a planting scheme for the garden of willow trees near the water and osiers, bamboos, maple and birch on the upper lawn.

Torrent House was published in a number of magazines and journals. With the exception of a couple of minor variations, the articles that appeared in *House Builder and Estate Developer* and *Interior Design* magazine are facsimiles and were probably authored by Wood or one of his team. Only the layouts of the two pieces and the photographs that accompanied them varied. *Leyland Journal*, the in-house magazine of the international automotive and engineering group discussed the house with somewhat greater originality. Even so, the article followed the model of other published texts quite closely, interpreting the project for readers in line with the wider ethos of Wood's practice.[14] *House and Garden* included the house in a special issue on 'the lure of conversions' in which it appeared alongside nine other examples.[15] The magazine appears to have been similarly dependent on Wood's publicity materials, emphasising the originality of the project and

Figure 9.4 Kenneth Wood. Torrent House, Hampton Court. The staircase, from a
leaflet produced by Peter Devenish.
Credit: Courtesy of Kenneth Wood.

its distinction from more routine domestic conversions: 'Anyone announ-
cing that he (or she) is thinking of converting an oast-house, barn or even a
lighthouse into a home-from-home causes no raised eyebrows. Converting
an electricity substation is quite another matter.'[16] Another contemporary
author recognised the functional principles on which the house had been
designed, but seemed rather unsure of its visual merits:

> The unusual design causes passing motorists to stare, but few realise
> that its shape is functional rather than a concession to any ultra-modern

Figure 9.5 Kenneth Wood. Oriel House, Haslemere.
Credit: RIBA Library Photographs Collection.

tastes held by the owners. The position of the house – just a few doors away from Hampton Lodge, a famous historic mansion and Paper House, once part of a dwelling owned by Sir Christopher Wren, makes its appearance even more incongruous.[17]

The Civic Trust, in contrast, recognised Wood's efforts to create a contextually appropriate building and welcomed the improvement that it had brought to the local environment. The citation for the Civic Trust award that was given to the house in 1965 stated:

> The rehabilitation of an abandoned electricity substation gets an award for imaginative (indeed witty) transformation of a typical subtopian eyesore. The resultant building is just right in character for the riverside world, trim, robust and yet with just that suggestion of a transitory nature that seems to pay the right respect to the permanence of the river.[18]

In the same year in which Torrent House won a Civic Trust award, the firm's remodelling and redecoration of a nearby solicitor's office, which showed a similar concern for the local context, was awarded a Civic Trust commendation:

Figure 9.6 Kenneth Wood. Oriel House, Haslemere. Interior.
Credit: Colin Westwood/RIBA Library Photographs Collection.

The Solicitor's Office is a sensitive, almost self-effacing, rehabilitation (and on the ground floor, transformation) of a good piece of folk-classical which had fallen into neglect. We are now so conditioned to accept the brutish savaging of the ground floors of so much good town building by assertive and insensitive shop fronts that it is a real pleasure to see an example which looks both of its own day and yet an integral part of an earlier building.[19]

Wood's respect for the architectural merit of existing buildings can also be seen in another large residential conversion, that of a semi-derelict mill, Felin Dawel, at Michaelston-Le-Pit in Wales. The house was designed for the conductor of the BBC Welsh Orchestra, John Carewe, his wife Rosemary Phillips and their children, and was the second house that he designed for the family. The first, Oriel House at Haslemere, was set in an acre of land on the north slope of a valley, on the site of an old orchard and was completed in 1963 (see Figures 9.5 and 9.6).[20] At Michaelston-Le-Pit, Wood partially reconstructed and substantially enlarged the old mill, using reclaimed materials in conjunction with a central timber-framed element designed to reconnect two older parts of the building.[21]

Notes

1 *House and Garden* (July 1966); *International Asbestos-Cement Review* (July 1966); *Leyland Journal* (October 1966); *London Transport Magazine* (January 1967); *MD Moebel Interior Design* (June 1969); *Interior Design* (November 1973); *House Builder* (June 1973); *The Evening News*, (20 January 1967); *25 New House Extensions and Improvements* (1972); *Interior Design* (November 1973).
2 The subject of converting larger family houses had been considered during the war. See, for example, H.V. Lanchester, 'Utilising obsolete houses', *The Builder* (16 June 1944), 477. On Mayorcas see, 'A house that grew', *House and Garden*, (Winter 1949): 40–1.
3 H. Dalton Clifford and R.E. Enthoven, *New Homes from Old Buildings* (London: Country Life Ltd, 1954), 9.
4 Clifford and Enthoven, *New Homes from Old Buildings*, 43.
5 Clifford and Enthoven, *New Homes from Old Buildings*, 67.
6 Clifford and Enthoven, *New Homes from Old Buildings*, 71.
7 Clifford and Enthoven, *New Homes from Old Buildings*, 71.
8 Clifford and Enthoven, *New Homes from Old Buildings*, 74.
9 Civic Amenities Act 1967, fifth impression (London: HMSO, 1968), 1.
10 The book was based on special issues of *The Architectural Review* that had focused on buildings of the functional tradition and Cantacuzino positioned it as 'a logical consequence' of J.M. Richards' book *The Functional Tradition in Early Industrial Buildings* (1958). Sherban Cantacuzino, *New Uses For Old Buildings* (London: Architectural Press, 1975). For a detailed review of literature in this area see B. Plevoets amd K. Van Cleempoel, 'Adaptive reuse as a strategy towards conservation of cultural heritage: a literature review', in C.A. Brebbia and L. Binda, eds. *Structural Studies, Repairs and Maintenance of Heritage Architecture XII*, (Southampton, UK: WIT Press, 2011), 155–64.
11 Cantacuzino, *New Uses For Old Buildings*, ix.
12 Cantacuzino, *New Uses For Old Buildings*, ix.
13 'There is an electricity sub-station in their house', undated newspaper clipping, KWPP.
14 'Living in a trolley bus sub-station', *Leyland Journal* (October 1966): 434–5.
15 The issue featured converted cottages, terraced houses and a mews house and conversions of a barn and a New York apartment. *House and Garden* (July 1966): 27–49.
16 'A London Transport sub-station becomes a Thames-side home', *House and Garden* (July 1966): 48.
17 'There is an electricity sub-station in their house', undated newspaper clipping, KWPP.
18 Civic Trust Awards Brochure, 1965, 80. KWPP.
19 Civic Trust Awards Brochure, 1965, 80. KWPP.
20 The oriel windows, a major feature of its design, lent a sense of spaciousness to four relatively small bedrooms on the first floor, while shading the windows of the living area below in hot weather.
21 The plan was developed to give an open southerly or westerly aspect to the main rooms, while solid walls to the north and east sides protected the house from winter winds. See, 'New life for a derelict mill house', *Stone Industries 6/3* (May–June 1971): 31; 'Conversion: Old Mill Glamorgan', *Interior Design* (May 1973): 333.

10 House for art
Picker House, 1968

Picker House at Kingston upon Thames was designed for Stanley Picker, a successful businessman and art collector, and was inspired by Picker's desire to create an architecturally distinctive setting in which to display his collection of modern and contemporary painting and sculpture (see Figure 10.1).[1] Although this house was a major commission for Wood, who developed a close working relationship with his client, Picker valued his domestic privacy and did not allow the publication of the project. There is, therefore, no record of the reception of the house at the time of its completion.

Largely unaltered since Picker took up residence in 1968, it is today significant as an exceptional British example of a late 1960s luxury house and interior.[2] The size of the project and requirements of his client were such that Wood worked in a coordinating role, designing the house and exterior landscaping and developing a schedule of loose furnishings and fittings for the interior. These were which were then completed in collaboration with Victor Shanley from Clifton Nurseries and with staff from Terence Conran's two companies, Conran Contracts and Conran Design Group.[3] Wood evolved his design for the house and its exterior spaces in response to Picker's living requirements, his art collection, and his aspirations for its display.

Picker grew up in New York and after graduating from Harvard Business School was sent to London to work in his family's pharmaceutical business, which he developed with great success from the late 1930s. In 1964 the firm was incorporated into the much larger Gala Cosmetic Group, with Picker as its Chairman and Managing Director. In the same year Picker began to look for an architect to fulfil his ambition of creating a house for art. Wood's firm was recommended to him by the commercial architects to Gala and was also put forward on a list of potential practices supplied by the Royal Institute of British Architects.

When Picker initially contacted Wood to discuss his plans he was unsure where to build and had two Surrey sites in mind, one at Weybridge, on which he had an option to buy, and another at Kingston, which he had purchased in 1957.[4] Wood favoured the plot at Kingston, which was steeply sloping and excited him with the potential that it offered to design a house

Figure 10.1 Kenneth Wood. The Picker House entrance court, *c.* 1968.
Credit: Courtesy of Kenneth Wood.

of individuality and interest.[5] Picker's initial brief to Wood was verbal. His chief requirements were for a house of 'generous scale' and 'imaginative design' with a living area large enough to host large social gatherings, suitable acoustics for the enjoyment of live and recorded music, adequate accommodation for guests and for two live-in staff.[6] His ambition was to create an informal setting in which to enjoy his art collection.

Plans for the house evolved over the course of visits that Wood made to Picker's home at Bathurst Mews in central London, to look at some of the art works that he already owned, and to the Gala factory and offices near Kingston, where part of his collection was displayed in the interior and grounds. Wood's final design for the house provided 4,975 square feet of living space at a cost of around £81,000. The house was completed between December 1966 and October 1968, when Picker took up residence. Wood was initially assisted on the project by Robert Jones and later by his principal architectural assistant, Martin Warne, who was responsible for the detailing.

The house was planned on two levels, to follow the natural fall of the land. Wood developed a ten foot structural and elevational grid, which he intended to 'impose a simple discipline upon the design, allowing the quality

Figure 10.2 Kenneth Wood. View of the Picker House from the terrace, *c.* 1968.
Credit: Courtesy of Kenneth Wood.

of materials and the works of art to make their statements without competi-tion'.[7] The house was designed with a load-bearing masonry podium and a glazed upper level of laminated post-and-beam construction. The timber and glass component was on a scale that was unusual in British domestic architec-ture at the time.[8] The deeply projecting eaves and the rain chains that Wood incorporated to draw water from the roof reveal an oriental influence, which is also evident in the layout and planting of the garden (see Figure 10.2).[9]

Wood developed the design around two courtyards. The first, an open courtyard, was created as an entrance and driveway. Picker acquired a Paul

Figure 10.3 Kenneth Wood. The Picker House internal courtyard, 1969.
Credit: Colin Westwood/RIBA Library Photographs Collection.

Mount sculpture, *Long Man* (1967), to stand beside the front door just before the house was completed (see Figure 10.1) and added *BB Project* (1973) by Anthony Twentyman in 1974. Overlooked by the entrance hall, a second inner courtyard separated the staff wing from Picker's private accommodation (see Figure 10.3). A hall, cloakroom, library and guest bedroom suite were located at entrance level and a large galleried living area, two further bedroom suites and a kitchen at lower level. Wood used the combination of courtyard and galleried elements to create visual and spatial interest within the modular plan (see Figure 10.4).

Figure 10.4 Kenneth Wood. View from the Picker House living area to the entrance
       gallery, *c.* 1968.
Photographer: Colin Westwood. Credit: Courtesy of The Stanley Picker Trust.

Wood's design for the house evolved in conjunction with discussions about
the presentation of Picker's art collection and the informal ambience that he
hoped to create. The entrance area was conceived as an important location
for the display of work. Early photographs of this part of the interior show
sculpture in the draught lobby and hall, an oil painting by Milton Avery,
Bedraggled Pigeon (1961), on one of the entrance hall walls, and on the
large slate wall adjoining the staircase a set of lithographs of Old Testament
subjects by Marc Chagall. Wood later commented on the arrangement of
the works: 'From the entrance one can progress, seeing part of the collection

Figure 10.5 Kenneth Wood. The Picker House master bedroom, *c.* 1968.
Photographer: Colin Westwood. Credit: Courtesy of The Stanley Picker Trust.

as background or studying it in detail.'[10] Certain fixed locations for the
display of work were evolved through the design process. Notable among
them was the large wall panel that Wood incorporated into the library as
hanging area for paintings. The panel, with suspended track lighting above,
was sited to diminish the visual effect of a wall of obscure glass, which
concealed an unattractive view over a flat roof toward a neighbouring property. It was chosen as a suitable place to display a work by Sidney Nolan,
Landscape (1968), which Picker wished to hang in a prominent position.[11]
In its library location the painting operated as a focal point and surrogate
for the absent view.

On the floor below, the main living area comprised formal and informal
dining zones and a conversation area, defined through the plan and the furniture groupings that Wood evolved in conjunction with the architectural
design of the building. The master bedroom suite – a dual-aspect bedroom,
a bathroom/sauna and a fitted dressing room – was located behind an unobtrusive door, angled beneath the overhanging library. A further bedroom
suite, off the formal dining zone, was similarly inconspicuous in its location,
lending to the central living space an illusory sense of autonomy. Wood
planned the interior to give garden outlooks to all of the major rooms: to

lawn and a woodland garden to the north, or to a terrace and water garden to the west. Acting as a buffer between Picker's private accommodation and the staff wing, a kitchen opened onto the formal dining area at one end and the informal dining area at the other.

Picker shared Wood's preference for the use of natural materials and the interior of the house was luxuriously fitted and decorated using a variety of timbers and other, primarily natural, floor and wall coverings. The muted colour palette for the walls was, according to Wood, 'to a degree intended to identify different functions but primarily restricted to colours and textures that would provide a relatively neutral background for paintings and sculpture'.[12] Correspondence between Wood and Picker reveals Picker's close attention to the visual relationship between his works and the architectural setting that Wood was creating for them. As the house was nearing completion he wrote to Wood of his new acquisition for the entrance court:

> I have verified the height etc. of the Paul Mount sculpture I have bought for the front of the house, 4'8" high, approximately 18" wide. Perhaps you would check this against the dimensions you talked about for the plinth and consider the total height in relation to the stone panel in front of which the sculpture goes.[13]

Wood's correspondence suggests that Picker was equally concerned with the appearance of works in conjunction with the wall coverings and finishes suggested for the interior. In a letter of 1967, he wrote to Picker: 'Small sample of grass paper G.421 enclosed for checking against Rodin bronze; larger sample can be obtained if necessary during my absence.'[14]

Terence Conran's two companies, Conran Contracts and Conran Design Group, played a major role in the design of the Picker House interior, introducing bright splashes of colour – orange velvet sofas, an apple green rug, and a set of Mies van der Rohe *Brno* dining chairs, upholstered in green – that complemented the vibrant colours of Picker's paintings, and in form and scale harmonised with Wood's design. In contrast to the cosmopolitan furniture suggestions of Conran's staff, Wood and his team selected fittings that were chiefly of British design and manufacture, subtly toned and sculptural in form.[15] Extensive built-in storage – coat cupboards in the hall, walk-in wardrobes in the guest rooms, a fitted dressing room off the master bedroom, a large sound fitment in the living area, and floor-to-ceiling cupboards in the formal dining zone, which read as wall panelling – concealed everyday objects from view and placed Picker's art works centre stage within a carefully orchestrated environment.

Wood's use of opposed planes in complementary materials structured visual experience of the interior in relation to Picker's collection and created associations between the house and the garden.[16] A major feature of the central living area was a low white shelf that extended the length of one of its walls. Wood designed it to display Picker's smaller works of sculpture and positioned it to draw the eye across the works to the terrace beyond. Outside, its visual counterpart, a low white wall of sculptural design, separated the staff lawn from Picker's garden.[17] The simple geometry of Wood's design created an understated setting in which to view Picker's collection. In turn, Picker's siting of works within the interior enhanced the experience of the building and its exterior landscape. Purchased in October 1968, the month in which Picker moved into the house, Enzo Plazotta's multiple portrait of Peter Ustinov was placed in a prominent corner location within the main living area, its central head facing diagonally across the interior and those below looking out to the terrace and water garden.[18] A corresponding location was chosen for the master bedroom, where Maurice Jadot's winged form, *Untitled 1/68* (1968), was placed on a low, glass table at one corner of the dual-aspect room. The 'axial links' forged through the location of works are of a type described by Henry and Lilian Stephenson in relation to professional design practice in their contemporary manual *Interior Design* (1964).[19]

Art was also incorporated into the fabric of Picker House, in the form of two specially commissioned windows. Early in 1966, Wood suggested that the wall between the entrance hall and the internal courtyard be 'handled as a solid/translucent screen with coloured glass' and enclosed two publications for his client's consideration, *Dowglaz Architectural Glass* by John Dowell and Sons and *Extended Uses of Glass*, which had just been published by the British Society of Master Glass Painters.[20] The idea of incorporating stained glass panels into the house was subsequently discussed, but a decision was deferred along with a number of items of additional expenditure.[21] Some months after Wood's original suggestion, when the house was nearing completion, Picker took up the idea of introducing stained glass into the design and asked Wood to approach two artists, Peter Tysoe and George Muller, on his behalf.[22] Tysoe, who was at that time Exhibitions Officer at the Crafts Council, designed two panels for the entrance hall.[23] These were situated to either side of a unit of clear glazing, framing the view to the internal courtyard (see Figure 10.6). A standing figure of *Eve* by Gillian Still was placed there soon after the house was completed (see Figure 10.3). The windows furthered the expression of Picker's artistic vision for the house and added to the privacy of the hall by screening it from the staff wing.

The second commissioned window, in dalles de verre and resin, was located in the main living area and played a similar role in constructing the interior as a private retreat (see Figure 10.7).[24] Muller, who was born in Zurich in 1940, studied stained glass in Germany under Adolf Valentin Saile and worked on the glass for the Metropolitan Cathedral of Christ the King in Liverpool (1967) in the studio of the prominent British stained glass artist Patrick Reyntiens.[25] The location of the window was suggested by the artist, to throw dappled blue light across the informal dining area.[26] Positioned at the boundary between Picker's private accommodation and the staff wing, it screened the main living space from the staff garden to its rear. A smaller

Figure 10.6 Kenneth Wood. The Picker House entrance hall showing windows designed by Peter Tysoe, *c.* 1968.
Photographer: Colin Westwood. Credit: Courtesy of The Stanley Picker Trust.

work by Muller, also in blue glass, was for some time located in the garden, amplifying the visual relationship between interior and exterior, house and collection.

Other material continuities between the house and garden reinforced those relationships in different ways. The white mosaic tiles that Wood selected for the master bathroom were used to line the upper pools of the water garden, including the large pool that was the setting for John Milne's sculpture *Leda* (see Figure 10.8). Although his previous projects had given

Figure 10.7 Kenneth Wood. View of the Picker House informal dining zone showing
a window designed by George Muller, *c.* 1968.

Photographer: Colin Westwood. Credit: Courtesy of The Stanley Picker Trust.

him limited opportunity to indulge his interest, Wood was alert to the visual
possibilities of locating art in the landscape. In the late 1950s he had visited
and photographed Henry Moore's garden, where he had observed the ways
in which planting could be used to create different experiences of sculpture
for the mobile viewer, a visit that had stayed in his mind. His early design
drawings for Picker House show his consideration of the placement of art
in relation to the building and exterior landscaping. A sculpture position
is indicated on the roof terrace adjoining the library and a strong vertical

Figure 10.8 Kenneth Wood. Sculpture on the upper terrace of the Picker House, 1969.
Credit: Colin Westwood/RIBA Library Photographs Collection.

form in the final pool of the water garden, for which Picker later acquired
Maurice Jadot's *Untitled 4:67* (1967), a tall vertical form in fibreglass, cast
from wood. As well as liaising with artists on Picker's behalf, Wood and his
staff designed sculpture plinths and settings for Picker's favoured works,
including an illuminated plinth for Elisabeth Frink's *Eagle* (1961) which
was placed in a prominent location on the terrace, where it could be seen
from the gallery and living area.[27]

Another means through which Picker's collection and the interior
and exterior of the house were related was through the use of light. The

Figure 10.9 Kenneth Wood. The Picker House gallery, *c*. 1977.
Credit: Courtesy of Kenneth Wood.

sophisticated electrical installation for the house included track, recessed and spot lighting in anodised aluminium or self-finished metal. The external lighting scheme was similarly extensive, incorporating down lights within the canopy to the front of the building; tree-mounted and shrub-mounted perimeter lights; pool lights for the water garden; spotlights and floodlights to illuminate sculpture on the main terrace; and flood lamps on the over-hanging eaves on the terrace and in the central courtyard. These were operated from a control panel within the master bedroom suite, allowing Picker to illuminate the exterior of the building and light the garden and art works within the grounds.

Wood's involvement with Picker continued after the house was completed and a number of other projects arose from it, including the remodelling of Picker's property in Bathurst Mews; the design of a paved garden and sculpture setting for his house in Spain; an interior design scheme for treatment rooms at Gala's Burlington Arcade showrooms in Piccadilly, which was never completed; and the design of a suite of New York offices for the Mary Quant Cosmetics brand. In 1974, as Picker was approaching retirement, he decided to build a private gallery in the grounds of the house to allow visitors increased access to his art collection, which had continued to expand in the six years since he had taken up residence.

Wood designed a pavilion gallery – a faceted circular structure on two levels, with a sunken, windowless floor to display and protect Picker's collection of paintings and a glazed upper floor to be used as a sculpture gallery (see Figure 10.9). The building was conceived as 'a jewel case for the exhibits' and placed the sculpture in a slightly elevated position where it could be viewed from the garden.[28] Formally, the gallery was conceived as a visual counterpoint to the rectilinear design of the house and Wood created a sense of continuity between the two spaces through his selection of materials and repetition of elements of the design of the main house, including the use of rain chains on the exterior. Located at the south-west corner of the site, the gallery was positioned to allow visitors direct access from the entrance courtyard.

The construction of the gallery required the replanning of the upper staff garden, and Clifton Nurseries were once more engaged to complete the landscaping and planting around the new building. Picker, who clearly trusted Wood's eye, asked him to prepare a hanging scheme for the picture gallery and a layout for the sculptures on the ground floor, designs for plinths, labels for the art works on both levels, and an entrance plate for the Stanley Picker Trust, which was completed by London firm, W. Perrin (Masonry) Ltd.[29] The gallery was completed between May 1976 and April 1977.

Notes

1 Little is known of Picker's collecting practices before the late 1950s, when he began buying from London art galleries. Picker's collecting and collection are examined by Fran Lloyd and Jonathan Black in their chapters on the Picker House sculpture and painting collection in Jonathan Black, David Falkner, Fiona Fisher, Fran Lloyd, Rebecca Preston and Penny Sparke, *The Picker House and Collection: A Late 1960s Home for Art and Design* (London: Philip Wilson, 2013).
2 Since Picker's death in 1982 the house has been managed by the Stanley Picker Trust, which Picker established to preserve his house and collection for future generations and to support the training of young artists.
3 These collaborations are explored in greater detail by Rebecca Preston and Penny Sparke in their chapters on the Picker House garden and interior in Black *et al.*, *The Picker House*.
4 H.M. Land Registry, Plan of Transfer dated 4 December 1957; letter from McKenna & Co. to the Stanley Picker Trust, 6 December 1984. SPT.
5 Wood believes that his enthusiasm for the Kingston site, which Picker also favoured, helped him win the commission. Kenneth Wood, in discussion with the author, March 2012.
6 'The Picker House (Number One)', 2. KWPP.
7 'The Picker House (Number One)', 3. KWPP.
8 Wood worked with William Willatts on the structural design.
9 See Rebecca Preston, 'The Picker Garden', in Black *et al.*, *The Picker House*, ch. 4.
10 'The Picker House (Number One)',3. KWPP.
11 Kenneth Wood, in discussion with the author, March 2012.

12 'The Picker House (Number One)', 5. KWPP.

13 S. Picker to K. Wood, 25 September 1968. SPPP.

14 K. Wood to S. Picker, 19 July 1967. SPPP.

15 Among them was the Adamsez 'Meridian One' range of sanitary ware, a light-weight vitreous china suite, designed by Alan Tye and Knud Holscher with Alan H. Adams in 1962. Its sculptural form and concealed plumbing won its designers first prize in an international competition organised by the International Union of Architects and the range was awarded a Design Centre Award in 1965.

16 One significant design development was a change to the design of the north-facing wall of the living area, which was originally planned with solid lower panels and later amended so that the glazing extended from floor to ceiling throughout. Letter from Kenneth Wood to Stanley Picker, 17 January 1966. SPPP.

17 The wall, which originally divided the staff garden from Picker's private garden, was removed from its original location when Picker added a private art gallery to the grounds in the 1970s.

18 As Fran Lloyd has indicated, from within the interior the portrait was positioned to direct attention back across the interior to Enzo Plazzotta's bronze *David*, which was displayed below the staircase. Fran Lloyd, 'The Picker House Sculpture Collection', in Black *et al.*, *The Picker House*, 145.

19 See, Henry and Lilian Stephenson, *Interior Design* (London: Studio Vista, 1964), 15 where the following is quoted:

> It will be noted that those buildings, which have a completeness and assimilation of design have these strong lines of interest linking various motifs, spaces, and vistas. Transcribed into three dimensions, it will be seen that the eye is 'turned' by particular emphasis on the 'axial centre' – a mosaic floor, a domed ceiling, a well-placed statue against a wall … These axial links help to relate the interior, to create vistas and points of interest, to turn the eye and therefore unconsciously direct the person. This directional punctuation is very important in interior design.

20 K. Wood to S. Picker, 17 January 1966. SPPP.

21 K. Wood to S. Picker, 5 December 1966. SPPP.

22 K. Wood to S. Picker, 22 March 1968. SPPP.

23 Peter Tysoe studied at Oxford School of Art and at Goldsmiths School of Art in London (1952–59) and worked as a Consultant Designer to Chance-Pilkington Glass Co. Ltd. (1968–70). See: http://www.petertysoe.net/index.html, accessed 30 December 2013.

24 At the time of the commission, Muller was represented by the John Whibley Gallery in Cork Street. J. Whibley to S. Picker, 1 March 1968. SPPP.

25 Biography of George Muller, The Sackville Gallery, East Grinstead, KWPP.

26 Kenneth Wood, in discussion with the author, March 2012.

27 K. Wood to S. Picker, 29 November 1968. SPPP.

28 Kenneth Wood, 'The Picker House (Number One)', 7. KWPP.

29 K. Wood to Mr Lawson at the Stanley Picker Trust, 27 October 1976. SPT.

Conclusion

The completion of Picker House in 1968 brought to a close the period in which Wood's firm was most involved in the design of private houses. It was also the year in which Royston Landau's *New Directions in British Architecture* considered the 'new spirit' that had begun to emerge in the work of Archigram, Cedric Price and others as they sought innovative architectural solutions to the problems of contemporary urban life and the design of modern dwellings. In his introductory chapter Landau pointed to the difficulty in accounting for the 'total scene' or 'context' of British architecture at the time, writing that it was

> neither simple to define, nor possible to completely describe, for even in the case of one solitary architect, he is likely to have his own network of interests and special set of information antennae, tuned in, perhaps to Tokyo or Carbondale, Illinois, so making this information exchange network part of his context.[1]

Without attempting to provide a precise account of Wood's context, this study has considered his 'network of interests' and the sources, values, practices and concerns that were most alive for him his approach to the design of the modern house matured.

Although the period in which Wood was involved in designing private houses was relatively short, it was one of significance for the development of domestic architecture in Britain, as architects reconsidered the work of pioneering avant-garde architects of the inter-war period and absorbed the more recent development of domestic architectural design in countries where progress had been less affected by the conflict. Wood, like many others, looked to Sweden and North America, where architects had continued to embrace the possibilities of new materials and technologies, while developing a more contextual approach to design that acknowledged more fully the domestic habits and preferences of users.

As Wood began to practice in the mid 1950s, greater consideration was being given to interior design and landscape architecture as distinct spheres of professional design practice and to the place of art in architecture. In

January 1956 *Architectural Design* magazine was redesigned and introduced its new look with a list of regular editorial additions:

> Of immediate interest is the fact that we shall allow more space from
> now on for Interior Design; and we shall shortly be starting up a 'land-
> scape architecture' feature which we hope will prove of great value. Also
> in response to repeated requests, we shall include regularly a detachable
> modern art page. This will take the form of a full page drawing by
> painters and sculptors who have something to say to architects, who
> will thus become acquainted with artists whose work they may like to
> use in their buildings.[2]

Wood was receptive to developments in associated design fields and used
them to further his own work.

While it would be wrong to link Wood's practice too closely to the use
of timber – the commercial work of his firm tells a somewhat different
story – structural timber elements feature in a number of his designs for private houses; a choice determined by his desire to create spaces suitable for
long-term occupation and adaptation and to create a freer domestic plan. In
different contexts, Wood used brick and timber in combination to express
functional and conceptual distinctions between interior spaces – not only
service and living spaces, but also fixed and flexible living areas, which is
most in evidence in his design for Whitewood. Used to promote the use of
timber in North American domestic architecture and interiors of the 1950s,
the term 'Livability Unlimited' expressed the values of flexibility, adaptability and comfort that underpinned Wood's approach to the design of houses
for British clients. Through his writings on timber construction, Wood was
among those who helped shape attitudes toward its use in British architectural design.

A respect for history and locality is evident in Wood's designs for buildings of all types and was recognised on several occasions by the Civic Trust.
It is also reflected in his involvement in street improvement schemes. As he
indicated, in connection with the latter:

> Most streets have built up their character over a long period by an evo-
> lutionary process, using materials locally available and appropriate for
> their position and in accord with the style of the time. More recently,
> this process has been speeded up by the availability of a much wider
> range of materials and a resultant lessening in regional characteristics.
> This has resulted in the introduction of inappropriate materials and
> colours, some of which create an entirely discordant effect. It should
> therefore be a principle of a street improvement scheme to retain the
> natural materials as far as possible, for they have stood the test of time
> and acquired a patina of use which becomes part of the character of
> the area.[3]

Wood began to design private houses during a period of changing popular taste. As Robert Furneaux Jordan observed in 1959:

> This change in taste is obvious and need not here be dilated upon. It is all round us. The Welfare State, the environment of the modern school, the example of the best public housing, the BBC, women's magazines, COID, technical education and popular science, new ways of living, travel, Espresso bars, Penguins, Festival of Britain, etc., have all played their part. Such popular taste is of course very easily debased, commercialized, exploited and, indeed, created for base motives. It is then called 'contemporary'.[4]

Wood never embraced the fashionable values of the contemporary style, but sought to create environments of enduring practicality and appeal. One result of his austerity-period training was a career-long concern with economy and ease of maintenance, which is reflected in his choice and use of materials, his attention to their care and repair and the modest appearance of his buildings.[5]

The types of environments in which Wood worked – established suburbs of late nineteenth or early twentieth-century origin and spaces on the fringe of expanding commuter towns – have often been presented as sites of architectural compromise in which modernism failed to fulfil its early social and aesthetic ambitions. However, it is clear from Wood's work that those environments were also attractive to clients who aspired to something different from the standard suburban semi-detached home. The professional location of Wood's practice was similarly intermediate: too closely connected to London's architectural scene to be classed as provincial, yet operating within distinct local networks that centred on Kingston upon Thames. Those networks incorporated architects and clients, major employers, such as Gala Cosmetics and the Hawker Aircraft Company, staff and students of Kingston College of Art, and Kingston's most significant retailers of domestic goods, who not only provided a setting for the display of local architecture, but were also places to which Wood's clients turned in furnishing and decorating their homes.

Wood was exceptionally client focused and the expertise of his practice lay in providing bespoke architectural solutions rather than models for mass development. Those clients were culturally engaged, middle-class professionals who were moderately advanced in their domestic aspirations and tastes. Personal preferences, possessions and collections, memories, experiences of past domestic spaces, hobbies and the need for professional workspaces all played a part in shaping the houses that Wood and his colleagues designed for them. For some, the appeal of modern architecture lay in the convenience of a cleaner, lighter and more technologically efficient domestic environment, while for others the choice appears to have been more closely associated with the possibilities that modern architectural design offered for the creation of spaces in which to live or work in a different way. In responding to his clients' specific requirements, certain of Wood's houses point to possibilities for hybrid domestic spaces combining elements of the workshop, studio or performance space and to more adaptable environments to serve the needs of families at different stages in their lives.

The firm's private houses were fairly widely disseminated, appearing in diverse editorial environments, as entire projects and as dispersed fragments of interiors in articles on kitchens, bathrooms, space-saving, lighting and materials, among others. Within those publications, Wood's authorial voice is often clear, shaping the reception of his work and determining its location within a professional and historical context. In the 1930s, Gordon Russell was among those who likened the more open living spaces that began to emerge to 'a return of the mediæval hall ... the communal room used for all activities of the house in the fifteenth century'.[6] References to historical and vernacular forms continued to play a role in mediating modern architectural design in Britain and internationally after the Second World War; and in the case of Wood's use of half-timbering in conjunction with open-plan living spaces, allowed their modern forms to be situated within a national architectural history – a strategy that he employed in his own writing.

It is perhaps fitting that the final words in this book be his. When asked about his architectural philosophy in 2008, his response was brief, swift and in the form of an aspiration for the future that was rooted in the values that informed his practice throughout his career: 'I would really like to see all architecture being humane. I would like to see it respect but not copy tradition and I think it should be really open to all influences which could come under the heading of social responsibility.'[7]

Notes

1 Royston Landau, *New Directions in British Architecture* (London: Studio Vista, 1968): 13–14.
2 'Apologia', *AD* (January 1956): 1.
3 'Chertsey Street Improvement Scheme', 2–3. KWPP.
4 R. Furneaux Jordan, 'Span. The spec builder as patron of modern architecture', *AR* (February 1959): 108–20, on 110–11.
5 Clients were provided with extensive maintenance manuals for their new houses. These detailed the materials and finishes that had been used along with appropriate cleaning products and methods to be employed in their care.
6 Gordon Russell from the 3 May 1933 broadcast, 'The living-room and furniture', in *Documents: A Collection of Source Material on the Modern Movement* (Milton Keynes: The Open University Press, 1979): 67.
7 Kenneth Wood, in discussion with the author, October 2008.

Bibliography

Journals

Architect & Building News, *Architects' Journal*, *The Architectural Review*, *Architectural Design*, B.C. Wood, *The Builder*, *The Canadian Architect*, *Country Life*, *Concrete Quarterly*, *Decorative Art. The Studio Yearbook*, *Design*, *De La Rue Journal*, *New Estates Magazine*, *Financial Times*, *Good Housekeeping*, *Home*, *Homemaker*, *Homes and Gardens*, *House Beautiful*, *House and Garden*, *Ideal Home*, *Illustrated Carpenter and Builder*, *Interbuild*, *The Lady*, *New Homes*, *Popular Mechanics*, *RIBA Journal*, *Sunday Times*, *Surrey Comet*, *The Times*, *Votre Maison*, *Woman's Journal*, *Woman's Own*, *Wood*, *The New House*.

Books and articles

Aitchison, Matthew. 'Townscape: scope, scale and extent', *Journal of Architecture* 17/5 (2012): 621–42.

Aldington, Peter. 'Architecture and the landscape obligation', *The Journal of the Twentieth Century Society* (2000): 20–8.

Allen, Gordon. *The Smaller House of To-Day*. London: B.T. Batsford, 1926.

Atkinson, Paul. 'Do it yourself: democracy and design', *Journal of Design History* 10/1 (2006): 1–10.

Attfield, Judy. *Bringing Modernity Home: Writings on Popular Design and Material Culture*. Manchester: Manchester University Press, 2007.

Aynsley, Jeremy and Kate Forde, eds. *Design and the Modern Magazine*. Manchester: Manchester University Press, 2007.

Banham, Reyner. 'Ateliers d'artistes. Paris studio houses and the modern movement', *The Architectural Review* (August 1956): 75–84

Banham, Reyner. *The New Brutalism, Ethic or Aesthetic*. London: Architectural Press, 1966.

Beecher, Mary Anne. 'Promoting the "unit idea": manufactured kitchen cabinets (1900–50)', *APT Bulletin* 32, 2/3 (2001): 27–37.

Bentall, Rowan. *My Store of Memories*. London: W.H. Allen, 1974.

Benton, Charlotte. 'Le Corbusier: furniture and the interior', *Journal of Design History* 3, 2/3 (1990): 103–24.

Benton, Tim. *The Modernist Home*. London: V&A Publications, 2006.

Bertram, Anthony. *The House a Machine for Living In.* London: A. & C. Black, 1935.

Bingham, Neil. 'The houses of Patrick Gwynne', *The Journal of the Twentieth Century Society* (2000): 30–44.

Black, Jonathan, David Falkner, Fiona Fisher, Fran Lloyd, Rebecca Preston and Penny Sparke. *The Picker House and Collection: A Late 1960s Home for Art and Design.* London: Philip Wilson, 2012.

Black, Jonathan and Brenda Martin. *Dora Gordine: Sculptor, Artist, Designer.* London: Dorich House Museum in association with Philip Wilson, 2007.

Blomfield, Sir Reginald. *Modernismus.* London: Macmillan, 1934.

Boydell, Christine. 'Textiles in the modern house', *The Journal of the Twentieth Century Society* (1996): 52–64.

Breward, Christopher and Ghislaine Wood, eds. *British Design from 1948: Innovation in the Modern Age.* London: V&A Publishing, 2012.

Burman, W., M. Pleydell-Bouverie and M.I. Urquhart. *Housecraft.* London: Macmillan, 1954.

Burstow, Robert. 'The *Sculpture in the Home* Exhibitions: Reconstructing the Home and Family in Post-War Britain'. In Henry Moore Institute Essays on Sculpture 60. Leeds: The Henry Moore Institute, 2008.

Butt, Baseden. 'An ultra-modern weekend-cottage', *The New House* (October 1935).

Campbell, Louise. 'Patrons of the modern house', *The Journal of the Twentieth Century Society* (1996): 42–50.

Cantacuzino, Sherban. *New Uses For Old Buildings.* London: Architectural Press, 1975.

Carrington, Noel. *Colour and Pattern in the Home.* London: B.T. Batsford, 1954.

Casson, Hugh, ed. *Inscape: The Design of Interiors.* London: Architectural Press, 1968.

Castillo, Greg. *Cold War on the Home Front: The Soft Power of Midcentury Design.* Minneapolis: University of Minnesota Press, 2010.

Charlton, Susannah, Elain Harwood and Alan Powers. *British Modern: Architecture and Design in the 1930s.* London: Twentieth Century Society, 2007.

Cherry, Bridget and Nikolaus Pevsner. *The Buildings of England: London 2 South*, Harmondsworth: Penguin, 1983.

Cieraad, Irene. '"Out of my kitchen!" Architecture, gender and domestic efficiency', *The Journal of Architecture* 7 (Autumn 2002): 263–79.

Clifford, H. Dalton. *Plan Your Own Home Decoration.* London: Country Life, 1955.

Clifford, H. Dalton. *New Houses for Moderate Means.* London: Country Life, 1957.

Clifford, H. Dalton. *Houses for To-Day.* London: Country Life, 1963.

Clifford, H. Dalton and R.E. Enthoven. *New Homes from Old Buildings*, London: Country Life, 1954.

Cohen, David H. 'A history of the marketing of British Columbia softwood lumber', *The Forestry Chronicle* 70/5 (September/October 1994): 578–84.

Collier, Allan. 'The Trend House Program', *Journal for the Society of the Study of Architecture in Canada* (June 1995): 51–4.

Colomina, Beatriz. 'The Media House', *Assemblage* 27 (August 1995): 55–66.

Conekin, Becky, Frank Mort and Chris Waters, eds. *Moments of Modernity: Reconstructing Britain 1945–64.* London and New York: Rivers Oram Press, 1999.

Constantine, Stephen. 'Amateur gardening and popular recreation in 19th and 20th centuries', *Journal of Social History* 14/3 (1981): 387–406.

Crinson, Mark and Jules Lubbock. *Architecture. Art or Profession?* Manchester and New York: Manchester University Press/The Prince of Wales Institute of Architecture, 1994.

Crinson, Mark. 'Picturesque and Intransigent: "creative tension" and collaboration in the early house projects of Stirling and Gowan', *Architectural History* 50 (2007): 267–95.

Cullen, Gordon. *Townscape*. London: Architectural Press, 1961.

Curtis, William J.R. *Modern Architecture Since 1900*. London: Phaidon Press, 1982.

Dannatt, Trevor. *Modern Architecture in Britain*. London: B.T. Batsford, 1959.

Darling, Elizabeth, *Re-Forming Britain: Narratives of Modernity Before Reconstruction*, London: Routledge, 2007.

Darling, Elizabeth. *Wells Coates*. London: RIBA, 2012.

Dean, David. *Documents: A Collection of Source Material on the Modern Movement*. Milton Keynes: The Open University Press, 1979.

Dean, David. *The Thirties: Recalling the English Architectural Scene*. London: Trefoil Books, in association with The RIBA Drawings Collection, 1983.

Elder, A.C., Ian M. Thom, Rachelle Chinnery, Allan Collier, R.H. Hubbard, Sherry McKay and Scott Watson. *A Modern Life, Art and Design in British Columbia, 1945–60*. Vancouver: Arsenal Pulp Press/Vancouver Art Gallery, 2004.

Eleb, Monique. 'Modernity and modernisation in postwar France: the third type of house', *The Journal of Architecture* 9 (Winter 2004): 495–514.

Endres, Kathleen L. and Therese L. Lueck, eds. *Women's Periodicals in the United States: Consumer Magazines*. Westport, CT: Greenwood Press, 1995.

Fisher, Fiona, Trevor Keeble, Patricia Lara-Betancourt and Brenda Martin. *Performance, Fashion and the Modern Interior from the Victorians to Today*. Oxford: Berg, 2011.

Fraser, Murray with Joe Kerr. *Architecture and the 'Special Relationship': The American Influence on Post-war British Architecture*. London: Routledge, 2007.

Friedman, A. *The Grow Home*. Montreal: McGill-Queen's University Press, 2001.

Friedman, A. *The Adaptable House: Designing for Choice and Change*. New York: McGraw-Hill, 2002.

Friedman, Alice T. *American Glamour and the Evolution of Modern Architecture*. New Haven, CT: Yale Univesity Press, 2010.

Gellen, Martin. *Accessory Apartments in Single-Family Housing*. New Brunswick, NJ: Transaction Publishers, 2012.

Giedion, Sigfried. *Space, Time and Architecture: The Growth of a New Tradition*, 5th edn. Cambridge, MA: Harvard University Press, 2008.

Girouard, Mark. 'A house built for a musician', *Country Life* (7 June 1962): 1383–5.

Girouard, Mark. *Big Jim. The Life and Work of James Stirling*. London: Chatto & Windus, 1998.

Gloag, John, ed. *Design in Modern Life*. London: George Allen & Unwin, 1946.

Gradidge, Roderick. *The Surrey Style*. Kingston: The Surrey Historic Buildings Trust, 1991.

Greenhalgh, Paul, ed. *Modernism in Design*. London: Reaktion, 1997.

Gould, Jeremy. 'Gazetteer of modern houses in the United Kingdom and the Republic of Northern Ireland', *The Journal of the Twentieth Century Society* (1996): 111–28.

Hamnett, Chris. *Winners and Losers: Home Ownership in Modern Britain*. Taylor & Francis e-library, 2005.

Hanson, Julienne with Bill Hillier, Hillaire Graham and David Rosenberg. *Decoding Homes and Houses*. Cambridge: Cambridge University Press, 2003.

Harwood, Elain. 'Post-war landscape and public housing', *Garden History* 28/1 (Summer, 2000): 102–16.

Harwood, Elain. *England: A Guide to Post-War Listed Buildings*, 2nd rev. edn. London: Batsford, 2003.

Hann, M.A. and K. Powers. 'Tibor Reich: a textile designer working in Stratford', *Textile History* 40/2 (November 2009): 212–28.

Heynan, Hilde. 'What belongs to architecture? Avant-garde ideas in the modern movement', *The Journal of Architecture* 4 (Summer 1999): 129–47.

Holmes, C.A. *New Vision for Housing*. Abingdon, Oxon: Routledge, 2006.

Holmes, C.A. *Homes for Today and Tomorrow*. London: HMSO, 1961.

Isenstadt, Sandy. *The Modern American House: Spaciousness and Middle Class Identity*. Cambridge: Cambridge University Press, 2006.

Isenstadt, Sandy. 'Modern in the middle', *Perspecta* 36 (2005): 62–72.

Jackson, Andrew. 'Labour as leisure: the Mirror dinghy and DIY sailors', *Journal of Design History* 19/1 (2006): 57–67.

Jackson, Lesley. *'Contemporary': Architecture and Interiors of the 1950s*. London: Phaidon, 1994.

Jensen, Finn. *Modernist Semis and Terraces in England*. Farnham, UK and Burlington, VT: Ashgate, 2012.

Jordan, R. Furneaux. 'Building in timber: advantages and limitations', *The Times* (3 March 1936): 49.

Jordan, R. Furneaux. 'Span. The spec builder as patron of modern architecture', *AR* (February 1959): 108–20, on 112.

Kaplan, Wendy, ed. *Living in a Modern Way: California Design, 1930–65*. Cambridge, MA: MIT Press/Los Angeles County Museum of Art, 2011.

Kenyon, Arthur W. 'Designs for sale: small house competition reviewed', *The Builder* (11 September 1959): 187–9.

Kirkham, Pat. 'At home with California modern, 1945–65'. In Wendy Kaplan, ed. *Living in a Modern Way: California Design, 1930–65*. Cambridge, MA: MIT Press/Los Angeles County Museum of Art, 2011.

Kynaston, David. *Austerity Britain, 1945–51*. London: Bloomsbury, 2007.

Kynaston, David. *Family Britain, 1951–57*. London: Bloomsbury, 2010.

Kynaston, David. *Modernity Britain: Opening the Box, 1957–59*. London: Bloomsbury, 2013.

Lancaster, Osbert. *Pillar to Post: The Pocket Lamp of Architecture*. London: John Murray, 1939.

Landau, Royston. *New Directions in British Architecture*. London: Studio Vista, 1968.

Langhamer, Claire. 'The meanings of home in postwar Britain', *Journal of Contemporary History* 40/2 (April 2005): 341–62.

Leavitt, Sarah. *From Catharine Beecher to Martha Stewart: A Cultural History of Domestic Advice*. Chapel Hill, CA: University of North Carolina Press, 2002.

Le Corbusier. *Towards a New Architecture*. Translated by Frederick Etchells. London: Architectural Press, 1927.

Lees-Maffei, Grace. 'From service to self-service: advice literature as design discourse, 1920–70', *Journal of Design History* 14/3 (2001): 187–206.

Liscombe, R.W. *The New Spirit: Modern Architecture in Vancouver, 1938–63.* Vancouver: Douglas and MacIntyre/Canadian Centre for Architecture 1997.

Long, Christopher. *Josef Frank: Life and Work.* Chicago: University of Chicago Press, 2002.

Lucas, Edgar. 'Conversion to open plan', *Homemaker* 1/3 (May 1959): 304–5.

Lupton Ellen and Jane Murphy. 'Case study house: comfort and convenience', *Assemblage* 24, (August 1994): 86–93.

McGrath, Raymond. *Twentieth-Century Houses.* London: Faber & Faber, 1934.

McKay, Sherry. 'Western living, western homes', *Society for the Study of Architecture in Canada Bulletin* 14/3 (September 1989): 65–73.

Mattson, Helena and Sven-Olov Wallenstein, eds. *Swedish Modernism. Architecture, Consumption and the Welfare State.* London: Black Dog, 2010.

Melvin, Jeremy. *F.R.S. Yorke and the Evolution of English Modernism.* Chichester: Wiley-Academy, 2003.

Melvin Jeremy and David Allford. 'F.R.S. Yorke and the modern house', *Journal of the Twentieth Century Society* (1996): 28–40.

Mills, Edward D. *1946–53. The New Architecture in Great Britain.* London: The Standard Catalogue Co. 1953.

Mills, Edward D. *The Modern Church.* London: Architectural Press, 1959.

Moran, Joe. '"Subtopias of good intentions": everyday landscapes in postwar Britain', *Cultural and Social History*, 4/3 (September 2007): 401–21.

Moran, Joe. 'Early cultures of gentrification in London, 1955–80', *Journal of Urban History* 34/1 (November 2007): 101–21.

Morand, François C. *La Maison Suburbaine.* Paris: Editions Albert Morancé, 1961.

Nairn, Ian and Nikolaus Pevsner. *The Buildings of England: Surrey*, rev. edn. New Haven, CT and London: Yale University Press, 1971.

Nairn, Ian. *Modern Buildings in Britain.* London: London Transport, 1964.

Nelson George and Henry Wright. *Tomorrow's House: A Complete Guide for the Home-Builder.* London: Architectural Press, 1945.

Newton, Miranda H. *Architects' London Houses.* Oxford: Butterworth Architecture, 1992.

Oldenziel, Ruth and Karin Zachmann, eds. *Cold War Kitchen: Americanization, Technology, and European Users.* Cambridge, MA and London: MIT Press, 2009.

Oliver, Paul, Ian Davis and Ian Bentley. *Dunroamin: The Suburban Semi and its Enemies.* London: Barrie & Jenkins, 1981.

Overy, Paul. *Light, Air and Openness: Modern Architecture Between the Wars.* London: Thames & Hudson, 2007.

Penoyre, John and Jane Penoyre. *Houses in the Landscape: A Regional Study of Vernacular Building Styles in England and Wales.* London: Faber & Faber, 1978.

Peto, James and Donna Loveday, eds. *Modern Britain 1929–39.* London: The Design Museum, 1999.

Petty, Margaret Maile. 'Curtains and the soft architecture of the American postwar domestic environment', *Home Cultures* 9/1 (2012): 35–56.

Plevoets, B. and K. Van Cleempoel. 'Adaptive reuse as a strategy towards conservation of cultural heritage: a literature review'. In C.A. Brebbia and L. Binda eds.

Structural Studies, Repairs and Maintenance of Heritage Architecture XII. Southampton: WIT Press, 2011.

Powell, Kenneth. *Powell & Moya.* London: RIBA, 2009.

Powell, Kenneth. *Richard Rogers Complete Works,* London: Phaidon, 2000.

Powers, Alan. *The Twentieth Century House in Britain: From the Archives of Country Life.* London: Aurum Press, 2004.

Powers, Alan. *Modern: The Modern Movement in Britain.* London: Merrell, 2005.

Powers, Alan. *Aldington, Craig and Collinge.* Twentieth Century Architects. London: RIBA, 2009.

Pritchard, Jack. *View from a Long Chair, The Memoirs of Jack Pritchard.* London: Routledge and Kegan Paul, 1984.

Reilly, C.H. 'The war and architecture'. In C.G. Holme, ed. *Decorative Art: The Studio Yearbook, 1941.* London: The Studio, 1941.

Rice, Charles. *The Emergence of the Interior: Architecture, Modernity, Domesticity.* London: Routledge, 2007.

Richards, J.M. *An Introduction to Modern Architecture,* rev. edn. Harmondsworth: Penguin, 1962.

Robertson, Howard. 'Domestic architecture and the second great war'. In C.G. Holme, ed. *Decorative Art: The Studio Yearbook 1940.* London: The Studio, 1940.

Rowntree, Diana. *Interior Design: A Penguin Handbook.* Harmondsworth: Penguin, 1964.

Rudberg, Eva. *The Stockholm Exhibition 1930: Modernism's Breakthrough in Swedish Architecture.* Translated by Paul B. Austen and Frances Lucas. Sweden: Stockholmia, 1999.

Ryan, Deborah S. *The Ideal Home Through the 20th Century.* London: Hazar, 1997.

Sanders, Joel. 'Curtain wars. Architects, decorators, and the 20th-century domestic interior', *Harvard Design Magazine* 16 (Winter/Spring 2002): 14–20.

Schneider, Tatjana and Jeremy Till. *Flexible Housing.* London: Architectural Press, 2007.

Schwab, Gerhard. *Einfamilienhäuser 1–50.* Stuttgart: Deutsche Verlags-Anstalt, 1962.

Schwab, Gerhard. *Einfamilienhäuser 51–100.* Stuttgart: Deutsche Verlags-Anstalt, 1966.

Sim, Duncan. *British Housing Design.* Coventry and Harlow: Institute of Housing and Longman Group, 1992.

Simms, Barbara. ed. *Eric Lyons and Span.* London: RIBA, 2006.

Smiley, D. 'Making the modified modern', *Perspecta* 32 (2001): 39–54.

Smith, Elizabeth. A.T. *Case Study Houses, 1945–66. The California Impetus.* Cologne: Taschen, 2007.

Smith, Ryan, E. *Prefab Architecture: A Guide to Modular Design and Construction.* Hoboken, NJ: John Wiley & Sons, 2010.

Sparke, Penny. *The Modern Interior.* London: Reaktion Books, 2008.

Sparke, Penny. *As Long as It's Pink: The Sexual Politics of Taste.* Halifax: Press of the Nova Scotia College of Art and Design, 2010.

Sparke, Penny, Anne Massey, Trevor Keeble and Brenda Martin, eds. *Designing the Modern Interior from the Victorians to Today.* Oxford: Berg, 2009.

Stalder, Laurent. '"New brutalism", "topology" and "image": some remarks on the architectural debates in England, around 1950', *The Journal of Architecture* 13/3 (2008): 263–81.

Stephenson, Henry and Lilian Stephenson. *Interior Design*. London: Studio Vista, 1964.

Walker, Lynne. 'Home making: an architectural perspective', *Signs* 27/3 (Spring 2002): 823–35.

Ward, Mary and Neville Ward. *Home in the Twenties and Thirties*. London: Ian Allan, 1978.

Waymark, Janet. *Modern Garden Design: Innovation Since 1900*. London: Thames & Hudson, 2003.

Webb, Michael. *Architecture in Britain Today*. Feltham, Middlesex: Country Life, 1969.

Whiting, Penelope. *New Single-Storey Houses*. London: Architectural Press, 1966.

Whyte, William. 'The Englishness of English architecture: modernism and the making of a national international style, 1927–57', *Journal of British Studies* 48/2 (April 2009): 441–65.

Williams, Richard. 'Representing architecture: the British architectural press in the 1960s', *Journal of Design History* 9/4 (1996): 285–96.

Womersley, Peter. 'Architects' approach to architecture', *RIBA Journal* (May 1969): 190–6.

Wright, Virginia. *Modern Furniture in Canada, 1929–70*. Toronto: University of Toronto Press, 1997.

Yerbury, F.R. *Small Modern English Houses*. London: Victor Gollancz 1929.

York, Keith 'Frank Lloyd Wright's legacy in San Diego', *Save Our Heritage Organisation, Reflections Quarterly Newsletter*, 37/1 (2006). Available at http://sohosandiego.org/reflections/2006-1/wrightlegacy.htm, accessed 16 December 2012.

Yorke, F.R.S. *The Modern House*, rev. edn. London: Architectural Press, 1951.

Yorke, F.R.S. and Penelope Whiting. *The New Small House*, rev. edn. London: Architectural Press, 1954.

Index

Figures are shown with a page reference in *italics*.